Cities and Nature

Cities and Nature fosters high-quality multi-disciplinary research addressing the interface between cities and the natural environment. It provides a valuable source of relevant knowledge for researchers, planners and policy-makers. The series welcomes empirically based, cutting-edge and theoretical research in urban geography, urban planning, environmental planning, urban ecology, regional science and economics. It publishes peer-reviewed edited and authored volumes on topics dealing with the urban and the environment nexus, including: spatial dynamics of urban built areas, urban and peri-urban agriculture, urban greening and green infrastructure, environmental planning, urban forests, urban ecology, regional dynamics and landscape fragmentation.

More information about this series at http://www.springer.com/series/10068

Andrea Arcidiacono • Silvia Ronchi
Editors

Ecosystem Services and Green Infrastructure

Perspectives from Spatial Planning in Italy

Springer

Editors
Andrea Arcidiacono
Department of Architecture and
Urban Studies
Politecnico di Milano
Milano, Italy

Silvia Ronchi
Department of Architecture and
Urban Studies
Politecnico di Milano
Milano, Italy

ISSN 2520-8306 ISSN 2520-8314 (electronic)
Cities and Nature
ISBN 978-3-030-54344-0 ISBN 978-3-030-54345-7 (eBook)
https://doi.org/10.1007/978-3-030-54345-7

This Springer imprint is published by the registered company Springer Nature Switzerland AG.
The registered company address is: Gewerbestrasse 11, 6330 Cham, Switzerland

Foreword

The last few decades have seen a flurry of activity in conceptualizing nature with the development of well-established legislative and strategic frameworks for green growth. In addition to this, advanced scientific studies have been conducted on a variety of green concepts (including studies on the key role of green spaces in improving the quality of life). Despite these efforts, cities and urban regions are still losing their green areas. The green structures of several European cities might still be visible, but hectares of public and private green spaces are alarmingly diminishing (see, e.g., the cities of Oslo, Copenhagen, Vienna, Helsinki, Milan, and Stockholm).

From the planning perspective, several green concepts have been used to address a sustainable urban development. Green corridors, green belts, green structures, and green fingers are intended to control urban sprawl, protect nature, and provide recreational opportunities. More recently, the concepts of green infrastructure (GI) and ecosystem services (ES) have been developed within the framework of the ongoing scientific debate and European and National policy. Compared to the traditional green concepts, GI and ES provide a holistic approach which integrates the socio-political and environmental concerns of landscape with contemporary urban planning. GI and ES can also be used to address a variety of several planning mandates (e.g., sustainable urban development, people's well-being, and quality of life).[1]

Moreover, regardless of the increasing collaboration among city and regional planning departments, universities and research institutes, as well as regional and master plans, all of the above have struggled to acknowledge the importance and role of green spaces within urban development. There are several challenges and obstacles in effectively embedding green concepts into and prioritizing green practices in planning.

[1] Mell I, Allin S, Reimer M. & Wilker J. (2017) Strategic green infrastructure planning in Germany and the UK: A transnational evaluation of the evolution of urban greening policy and practice, *International Planning Studies*, 22:4, 333-349.

These impediments might be related to local socio-political dynamics, private interests in the urban development process, as well as a lack of expertise in the planning departments.

In this foreword, I highlight the following issues related to GI and ES as well as their implementation in planning. These issues have been theoretically and empirically analyzed in recent years (by analyzing several cases in Italy, Canada, Finland, and Norway) in these four areas of: i) *the transfer of advanced scientific knowledge of GI and ES into planning*; ii) *implementation of the concepts of GI and ES at different levels of planning* (moving from regional to local projects); iii) *rhetoric images of green cities*, which are related to the predominance of certain "green interests" in city developments (see, e.g., real estate developers) and the idealized representation of green by planners; and iv) *the model of urban growth that cities address and related impacts on green* (e.g., compact city vs urban sprawl).

These arguments, which are expanded upon below, help underpin the core rational of current and innovative studies in the European and Italian contexts which are presented in this book.

(i) The transfer of advanced scientific knowledge of GI and ES into planning strategies and practices. These emerging green concepts have been developed by experts from different scientific fields (e.g., biology, landscape architecture, and landscape ecology) and then transferred (or are being transferred) by policymakers and planners into the green plans of our urban regions and cities (which often do not have any legal status). The traditional expertise of planning and related fields is inadequate to understand the role and functions of GI and ES in urban development.[2] More interdisciplinary groups of practitioners are needed in the planning divisions of our cities and regions.

(ii) Implementation of the concepts of GI and ES at different levels of planning. It seems that GI and ES are often used as the conceptual framework when designing strategic scenarios at the regional and local scales.[3] However, the implementation of ES into current land-use planning (e.g., zoning and detailed plans) is still rather difficult. For example, the mapping of ES cannot be easily translated into current land-use maps by using the existing planning tools (and regulation). Moreover, at the level of the urban projects, we should recall that storm water management (which refers to the regulating ES), protection of nature (see supporting ES), as well as access and view to green spaces (see recreational ES) have already been acknowledged for decades by planners, but only recently have they been included in the ES conceptual framework. This can generate some misinterpretations of the concepts of ES and related purposes.

[2] Lahde E. & Di Marino M. (2018). Multidisciplinary collaboration and understanding of Green Infrastructure. Results from the cities of Tampere, Vantaa and Jyvaskyla (Finland). *Urban Forestry and Urban Greening* https://www.sciencedirect.com/science/article/pii/S161886671730420X

[3] Di Marino M., Tiitu M., Lapintie K., Viinikka A. and Kopperoinen L. (2019) 'Integrating green infrastructure and ecosystem services in land use planning. Results from two Finnish case studies'. *Land Use Policy,* 82: 643-656.

Furthermore, the administrative fragmentation and sectorial plans that characterize several contexts, which have been analyzed in this book, do not support the implementation of GI and ES at different levels of planning and across boundaries. This administrative fragmentation is more evident in the Italian context. Municipalities make land-use decisions through local development plans (local master plans and detailed plans), while regions, provinces, and metropolitan cities produce regional landscape plans and regional territorial plans, provincial territorial coordination plans, and strategic metropolitan plans, respectively. On one hand, each region provides its own regional framework law on land use and might extend it to implement GI and ES. On the other hand, the absence of a national framework law on land use does not allow the regions to create coherent policies for GI and ES. Nevertheless, in both Italian and European cities and regions, soft-planning tools and non-statutory planning might be seen as being relevant to supporting the incorporation of these concepts.

(iii) Rhetoric images of green cities have characterized the most recent urban development (at the city, regional, and building level). In urban development processes, there are different "green interests." Very often, the predominance of the private development itself has compromised the achievement of public planning objectives. Real estate developers have tended to emphasize the property looking onto a green view, which typically increases housing costs and property values. This approach has resulted in limiting urban densification around protected areas (based on given distances and potential impacts), while, in contrast, the other green areas have been constantly affected by urban development. These processes have also resulted in a persuasive storytelling and representation of green cities. Thus, green concepts have gained a metaphorical power in planning. The Dutch green heart, for example, "is a metaphor that masks the real physical features of the area"[4] which neither exists in plans nor does it form a homogenous unity. The risk is that GI and ES can also be used for greenwashing land-use projects with little ecological value.

(iv) The model of urban growth and impacts on green. Several cities have followed the model of a compact city in order to preserve arable lands and forests as well as their biodiversity from urban growth. In the last 10 years, the compact city has been considered the most environmentally sustainable option for urban form as well as public policy.[5] Although the compact city model can protect against greenfield development as well as help preserve nature in terms of forests and farmlands, it has exerted extreme pressure on urban nature, especially on urban green (both public and private). Considering this model of growth, for example, urban ES are really endangered, such as (1) recreation and creativity, (2) engagement with nature (and well-being), (3) human thermal comfort, and (4) air quality.

[4] Van Eeten N., & Roe E (2000) When Fiction Conveys Truth and Authority, *Journal of the American Planning Association*, 66:1, 58-67, p 61.

[5] Mouratidis, K. (2017). Is compact city livable? The impact of compact versus sprawled neighbourhoods on neighbourhood satisfaction. *Urban studies*, doi: https://doi.org/10.1177/0042098017729109

Thus, the green is not prioritized within urban development. The green spaces are often readily sacrificed as cities expand and develop.[6] This loss of green is often compensated for by the increasing share of green spaces in other urban areas (e.g., the mechanism of ecological compensations). We also see "green left-overs" in the built environment. These approaches confirm that the provision and quality of green generally hold a low priority in urban development. The green is diminishing in our urban regions, and to date, this trend has not been reversed. The concepts of GI and ES are rather new in planning; therefore, we do not yet have a broad perspective and understanding of the possible impacts of this trend.

In the last few years, although there has been a consolidation of the concepts of GI and ES within the scientific debate and policy frameworks, the assimilation and adaptation of both concepts in spatial planning (both research and practice) is still rather new. Understanding GI and ES requires new expertise, interdisciplinary approaches in planning, and probably new planning tools. It is hoped that the European and Italian cases in this book will show a comprehensive overview of the potential of both concepts and possible implementation.

Department of Urban and Regional Planning Mina Di Marino
NMBU, As, Norway
e-mail: mina.di.marino@nmbu.no

[6]Di Marino M., Niemelä J. & Lapintie K. (2018). Urban nature for land use planning. *Urbanistica*, 159: 94-102.

Contents

1 **Challenges for Contemporary Spatial Planning in Italy. Towards a New Paradigm** ... 1
Andrea Arcidiacono and Silvia Ronchi

2 **Urban Green Infrastructure: Opportunities and Challenges at the European Scale** ... 17
Grazia Zulian, Julie Raynal, Rayka Hauser, and Joachim Maes

3 **The Green City: From a Vision to a Concept from National to European Perspectives** ... 29
Jürgen H. Breuste

4 **The Inertial Forces of Ecological Planning: How Planning Resists Conceptual Change** ... 45
Kimmo Lapintie

5 **The Project of the Green Infrastructure in Lombardy Region. A Resilient Spatial Structure for the Landscape Plan** ... 59
Andrea Arcidiacono and Silvia Ronchi

6 **A Green Infrastructure in the Guidelines to Limit Land Consumption of the Friuli Venezia Giulia Regional Landscape Plan** ... 73
Elisabetta Peccol, Mirko Pellegrini, and Mauro Pascolini

7 **The Landscape Planning and the Green Infrastructure in Campania Region** ... 87
Emanuela Coppola

8 **The Green Infrastructure Instrument for the Metropolitan Area of Naples: Experimentations Through Local Planning** ... 101
Francesco Domenico Moccia and Antonia Arena

9 Green Infrastructure and Local Planning Processes: A Study Concerning the Metropolitan Context of Cagliari 113
Sabrina Lai, Federica Leone, and Corrado Zoppi

10 Ecosystem Services Integration into Local Policies and Strategies in the City of Bologna: Analysis of the State of the Art and Recommendations for Future Development 127
Claudia de Luca, John Martin, and Simona Tondelli

11 The New Urban Plan of Rescaldina Municipality. An Experience for Improving Ecosystem Services Provision 141
Silvia Ronchi, Andrea Arcidiacono, and Laura Pogliani

12 Identifying Ecosystem Service Hotspots to Support Urban Planning in Trento 153
Davide Geneletti and Chiara Cortinovis

13 Mapping Ecosystem Services, Disservices, and Ecological Requirements to Enhance Urban Forest Planning and Management in Padova 167
Chiara Cortinovis, Claudia Alzetta, and Davide Geneletti

14 Messina. Green and Blue Infrastructures for the Re-urbanisation of the City 181
Carlo Gasparrini and Anna Terracciano

15 Green Texture: Nature and Reuse in the Prato Operative Plan Legislation 201
Francesco Caporaso, Pamela Bracciotti, and Antonella Perretta

16 Green Infrastructure and Landscape Planning in a Sustainable and Resilient Perspective 213
Angioletta Voghera and Benedetta Giudice

17 Lessons from Italian Experiences: Bottlenecks, New Challenges and Opportunities 225
Silvia Ronchi and Andrea Arcidiacono

Index 237

About the Editors

Andrea Arcidiacono MSc., Arch., Ph.D., in Urban Regional and Environmental Planning at the Politecnico di Milano. Associate Professor at the DAStU (Dep. of Architecture and Urban Studies), Politecnico di Milano. He is Director of the LabPPTE (Landscape Plans Territories Ecosystems Lab) and member of the scientific board of the CRCS (Land Take Research Centre). Since 2016, Andrea is Vice President of the National Institute of Urban Planning (INU). He is also a Member of the Editorial Board of the national journal *Urbanistica*. His research focuses on landscape and urban planning, open spaces and green and blue infrastructure design, ecosystem analysis and nature-based solutions for spatial planning, and policies for land take limitation. Currently, Andrea is the DAStU Principal Investigator in the program LIFE 2014–2020 "European Programme for the Environment and Climate Action" (LIFE 2017), "'SOIL4LIFE," and Project Leader (DAStU-POLIMI) for studies and researches to support the revision of the Lombardy Regional Landscape Plan. He has been a member of the Scientific Committee for the Town Plan of the City of Milan (2012). Andrea has been a consultant of several public Administrations and designer of town plans and detail plans. He is (co)author of over 120 scientific publications.

Silvia Ronchi MSc. Urban Planning, Ph.D., in Urban Planning, Design and Policy at the Politecnico di Milano. Untenured Researcher and Assistant Professor at the DAStU (Dep. of Architecture and Urban Studies), Politecnico di Milano and member of the LabPPTE (Landscape Plans Territories Ecosystems Lab, Politecnico di Milano). Since 2007, Silvia is member of the Scientific Board of the CRCS (Land Take Research Centre) and Editor of the CRCS annual national reports. She is specialized in Geographic Information System for Spatial planning. From 2015 to 2018, she collaborated with the Joint Research Centre of the European Commission (Territorial Development Unit – B3). She is a member of the Ecosystem Services Partnership (ESP) and the International Association for Landscape Ecology (IALE). Her research focuses on Ecosystem services, landscape and urban planning, environmental assessment, green and blue infrastructures design. She has been a consultant of several Public administrations and research centres for the Strategic Environmental Assessment of General Town plans and sustainable development projects (ecological connection, landscape studies and natural capital). She is (co) author of over 50 scientific publications.

Chapter 1
Challenges for Contemporary Spatial Planning in Italy. Towards a New Paradigm

Andrea Arcidiacono and Silvia Ronchi

Abstract The new environmental, ecological and social emergencies affecting the contemporary city and territory of the "Anthropocene era" have increasingly intense impacts on human well-being and quality of urban life. Emergencies, closely related to regional anthropisation processes, concern issues of adaptation to climate change, risk prevention and food security. Responding to these challenges requires a shift in strategies and urban design models. In Italy, traditional planning models still prevail, mainly oriented towards governing processes of urban growth and improving regional infrastructures, which strongly affect the availability of natural resources. Even recent planning experiences, focused mostly on the governance of urban redevelopment processes, have been unable to reduce the persisting intensity of urbanisation processes or trigger broader regeneration effects within the increasingly less efficient and less liveable urban fabrics of the built-up city. Nowadays, it is necessary to redefine the territorial governance agenda and experiment with a new urban planning paradigm which can address the re-urbanisation of the contemporary city in an ecologically oriented and socially cohesive perspective, guaranteeing the well-being and the quality of citizens' lives through a robust reconstruction of the urban natural capital.

Keywords Planning perspective · Urbanisation and land take limitation · Climate change adaptation · Ecosystem services · Green and blue infrastructures · Nature-based solutions · Resilient regeneration · Human well-being · Contemporary cities

---

A. Arcidiacono (✉) · S. Ronchi
Department of Architecture and Urban Studies, Politecnico di Milano, Milano, Italy
e-mail: andrea.arcidiacono@polimi.it

A. Arcidiacono, S. Ronchi (eds.), *Ecosystem Services and Green Infrastructure*, Cities and Nature, https://doi.org/10.1007/978-3-030-54345-7_1

1.1 Introduction

In recent decades, the intensity of anthropisation processes has had an irreversible impact on the availability of natural resources (water, soil, and air) with direct effects on the production of food and raw materials, on hydrogeological stability, and, more generally, on climate change.

In about a century, commonly known as "Anthropocene," which began with the late nineteenth-century modern era, human action has started to produce radical and irreversible changes on the biophysical composition of our planet (Crutzen 1998, 2005). Anthropisation processes are among the most significant determinants of the planet's environmental and ecological crisis (UN 2019a). If the process of growth continues at its current intensity, urban areas will contain about 70% of the world population by 2050, and the land occupied by cities in the developing world will triple (UN 2019b).

The physical features and environmental conditions of these urban conglomerations are profoundly different in various areas of the planet. In Europe, the urbanisation process is more advanced than in other regions of the Earth. In Italy, over 70% of the population already lives in urbanised areas with similar structure – compact central urban areas with important historical features; dense and compact expansion fabrics from late nineteenth-century and early twentieth-century development with a mix of residential, productive and craft activities; large peripheral and peri-urban mono-functional areas, where large industrial settlements and new residential districts stand side by side, developed during the second half of the twentieth century, during the phase of maximum settlement and employment development of European cities. The mass development phase coincided with the maximum intensity of "land take" processes, which are mainly due to the urban transformation of agricultural soils. In Italy, from the 1980s, this long period of urban growth demonstrated more complex and articulated aspects. On the one hand, there was the onset of intensive reconversion processes of large disused urban industrial areas or areas linked to the abandonment and under-utilisation of primary urban services (customs, slaughterhouses, railway yards and, more recently, barracks). On the other hand, the going on of urban growth, which despite being apparently less intense compared to the dynamics of the decades of post-War mass expansion, started to increasingly affect peri-urban and suburban areas connoted by the ever-increasing spread of low-density settlements determining a huge peri-urbanisation process as one of the most unsustainable forms of urban development. In a situation of weakness and substantial inadequacy of the intermediate levels of territorial government (provincial and supra-local initially, and now metropolitan authorities), sprawl processes have provoked peri-urban growth through intense land take and soil sealing with high landscape fragmentation. This is due to the prevalence of a horizontal growth of settlement characterised by low-density housing models and high land occupation (Brueckner 2001; EEA 2006), and the consequent emergence of a mobility system based on the use of individual vehicles, necessary because of a lack of connection between public transport planning with regional development strategies; and simultaneously supported by the spreading of a capillary road network. Sprawling urban development caused diffuse anthropisation of rural environments which were characterised by high landscape value and significant production capacity, and it profoundly affected the continuity and quality of ecological and environmental connectivity of the peri-urban area.

This urban development dynamic was underestimated by Italian urban planners who believed that the new season of "urban transformation," focused on the reuse and redevelopment of large abandoned urban areas, would marginalise expansive growth. During the 1990s and for a large part of the first decade of the 2000s, in Europe and Italy, the reuse of abandoned and underused sites located in central urban areas has coexisted with the strengthening of new land take phenomena. In Italy, over about 30 years (1989–2018), more than 7700 km^2 of agricultural and natural soils have been urbanised (ISPRA 2016, 2018), with an annual average of more than 260 km^2 (almost one and a half times the extension of the city of Milan and roughly double its urbanised area). This intense land take process led the country to have an anthropised surface area of more than 7.5% of its territory (ISPRA 2018). This phenomenon was not only due to the prevalence of low-density scattered settlement patterns and the spread of new types of production and work activities (shopping malls and logistics hubs) but mainly to the persistence of economic and real estate pressure produced by the incidence of the urban rent.

Land take and soil sealing remain among the leading causes of soil degradation processes across Europe (Ronchi et al. 2019): with direct impacts on the reduction and deterioration of one of the primary natural resources, that provides fundamental ecosystem services for quality of human life and well-being; with effects on food production, air quality, water regulation, hydrogeological stability and more generally on climate change. These effects influence the salubrity of our cities and the health of citizens, who are increasingly exposed to diseases linked to the intensity of urbanisation phenomena and soil sealing. The effects of heatwaves in central Europe are one of the main causes of death during the four summer months (EEA 2016; Geneletti et al. 2020); the record temperatures registered in different parts of Europe in 2013, 2014 and 2015 led to an exceptional increase of mortality (Gasparrini et al. 2015). In 2015, the summer heatwave caused more than 3000 deaths in France (EEA 2017). In the last 20 years, the European Commission has widely reported the risks and impacts connected to the persistence of soil sealing and land take processes in Europe and has outlined the main strategies to be implemented to limit the intensity of these phenomena (European Commission 2002, 2006a, 2012) towards the goal of "no net land take" by 2050 (European Commission 2016). Nevertheless, the European Commission has failed to approve the Soil Framework Directive (European Commission 2006b). This act would have strengthened the legislative action of the Member States (Ronchi et al. 2019), but it was withdrawn in 2014 at the wishes of some of the leading Member States. The reasons for the opposition include the subsidiarity and proportionality principles, the estimated costs, the administrative burden and existing national legislation on soil that was not considered as aligned with the incoming proposal (Glæsner et al. 2014).

Many European States activated public policies and legislative measures to reduce land take and soil sealing and support urban regeneration for new environmental and social liveability, acting both on quantitative limitation and on fiscal policies. In Italy, throughout the first decade of the 2000s, the national annual average of land take intensity exceeded 60 hectares per day (ISPRA 2016). This trend was more intense in some regions, such as Lombardy and Veneto, often with a significant impact on landscapes of outstanding environmental value. In recent years, land take processes, while remaining intense, registered a considerable reduction. In 2018, the annual average was less than 15 hectares per day, with an overall increase in artificial surface areas of 51 km^2 compared to a total amount of national urbanised area of 23,033 km^2 (ISPRA 2019). This reduction was not so much attributable to national or regional legislative measures, which are still absent or not yet implemented. All national Governments that have, since 2010, sought to approve, with differing determination, a draft law to limit land take and to incentivise urban regeneration, failed. Regionally, new legislations approved in recent years have been partly contradictory and only partially applied in planning tools. Instead, the land take process decrease was mainly the global effect of the economic crisis that also affected the construction sector and the real estate market. However, in the face of a factual reduction of urbanisation processes, urban plans in Italy continue to propose a development model still mainly oriented to urban growth dynamics, which are incoherent with demographic and employment trends – an approach purely based upon speculative real estate and financial logics that will perpetuate for a long time a new land take process (Arcidiacono 2015). This planning model is still far from dealing effectively with the current environmental and ecological emergencies. These are issues directly concerning the definition of innovative strategies in urban planning, oriented towards the construction of adaptive and resilient actions able to respond to increasingly intense territorial risks and reduce the ongoing effects of climate change.

1.2 New Priorities for Urban Planning: Redefining the "Common Interest"

Despite the intense land take processes, in Italy, it seems still arduous to approve a legislative reform that supports spatial planning aimed at addressing land take reduction and promoting urban regeneration interventions – a framework law that defines principles and planning priorities for a resilient approach in the designing of the contemporary city and to contrast climate change through adaptive planning solutions (Arcidiacono 2015). The planning models currently used in Italy are still traditional, driven by logics of urban and infrastructures growth, often divorced from demographic or employment requirements. Development strategies and choices are made by the administrative municipal level, within a planning system in which diverse territorial planning levels (provincial, metropolitan and regional) have never had the strength or efficacy to guide, coordinate or influence decisions on local land-use planning. Nowadays, the forecasted urban transformations that threaten soil and ecosystem services are defined in local urban plans. The adoption of supra-local scale can reduce and mitigate these impacts where the design of environmental and ecological networks and the construction of green belts can be effectively and coherently planned, and the future development decisions calibrated according to the actual forecast of population and employment growth.

Quantitative limitation of land take must be introduced, applying legislative acts, planning and land-use conformation tools at the most appropriate territorial levels, and continuing to monitor the extent and the intensity of the processes. Nevertheless, mapping the land take process or introducing normative rules for its quantitative restriction is not sufficient; it is fundamental to introduce a qualitative assessment approach that considers not only the amount of soil surface loss but also soil quality and the ecosystem services provision to evaluate and select appropriate design strategies aiming to enhance ecosystem capacity and related benefits, which are crucial for quality of life and human well-being (Millennium Ecosystem Assessment 2005). The soil ecosystem services directly concern air quality, water filtering and regulation, food production, landscape quality, cultural and aesthetic historical values, and deeply affect climate change and environmental risks that are increasingly impactful and perceivable in urban contexts. Anthropisation and soil sealing due to urbanisation processes involve a reduction, and often a zeroing, of ecosystem services provision, with significant impacts on citizens' quality of life and health conditions (Dodge et al. 2012; Shekhar et al. 2019). In urban and peri-urban areas, where the contribution of regulating ecosystem services is most relevant, the quality and improvement of health and well-being conditions are connected to the physical and morphological features of the built environment, and the availability and condition of natural and green open spaces (WHO 2019).

A radical update of the traditional urban planning paradigm is needed for designing the contemporary city to face with these emerging pressure conditions and the urgency to provide adaptive and resilient responses to climate change (Arcidiacono et al. 2018a); an innovation of spatial planning model that places ecological and environmental issues at the centre of the design and planning action to conserve and strengthen the provision of those ecosystem services on which life quality and inhabitants' well-being depend. This perspective requires a full-fledged update of the "common and public goals" for urban planning. Fifty years after the Henri Lefebvre essay on the "Right to the city" (Lefevre 1968), the needs, rights and desires of citizens have changed, including social and distributive equality terms and requirements of wellness and liveability of cities. In Italy, "Planning standards" (introduced into the Italian legislation at the end of the 1960s, with Inter-Ministerial decree no. 1444/68) have traditionally guaranteed adequate conditions of local welfare, introducing a mandatory minimum supplies of public spaces (parks and socialisation spaces) and services (school, health and cultural), as availability and spatial configuration, to provide an acceptable level of urban liveability (Giaimo 2019). Today, these facilities are still fundamental for structuring the fixed capital of the "public city." A large part of Italian cities' quality of life depends upon assigning different functions and values to urban spaces, indiscriminately offering minimum supplies of public areas and services to everyone, despite the issues of the social division of the space not being addressed. However, it becomes inevitable to expand and redefine the boundaries of urban plan's "common and public interest", introducing notions of social, ecological and environmental performances, related to the complex and widespread forms of the contemporary city. A process of urban planning innovation that poses articulated challenges requires different levels of experimentation (Ronchi et al. 2020). On the one hand, this involves redefining the spatial plan and urban structure framework around the design of public space, and environmental and usage networks of the contemporary city (e.g. green and blue infrastructures); on the other hand, this involves introducing qualitative and performance parameters and indicators, which should be adequate to verify the sustainability of the plan's strategies in a perspective of adaptation and resilience (Schewenius et al. 2014) to environmental changes, complexity of physical and social contexts and new demand for welfare. These performance standards are relevant for updating the traditional urban plan features and contents to be renewed in its processes and tools, but remaining stable in its goals, to protect everyone's interests and defend the quality of spaces where people live.

1.3 Ecosystem Services for Supporting a New Spatial Planning Paradigm

The new environmental and ecological emergencies require redefinition the concept of "common and public interest" in a broader categorisation of services provided to citizens for their well-being, and in a qualitative perspective of performance and resilience in defining land uses. According to this, the introduction of a different planning paradigm finds a relevant contribution from the ecosystem services approach.

The increasing relevance and dissemination of ecosystem functions and services in environmental and soil science researches (Burkhard et al. 2012, 2013; Haase et al. 2014b) can provide an effective contribution to spatial and landscape planning, at reconfiguring the plan's environmental and ecological structure, at selecting land-use strategies for defining regeneration and re-urbanisation solutions for the contemporary city and landscape (Cortinovis and Geneletti 2019). The mapping and evaluation of ecosystem services (Maes et al. 2016), defined in literature as multiple benefits provided by ecosystems to humanity (Millenium Ecosystem Assessment 2005; Haines-Young and Potschin 2013), have become an increasingly accurate and investigated research subject. This approach can constitute a fundamental contribution to improving the decision-making processes for spatial planning. The measurement and monitoring of ecosystem services can facilitate a comparative and dynamic assessment of the effects determined by alternative scenarios of transformation and development of land use on urban quality and support the identification of multiple common performance objectives in deciding "where to put things" (Polasky et al. 2008). In this way, the ecosystem approach acts as a model to interpret and address the collective contemporary city "needs," as it assumes the perspective of beneficiaries, who are the service recipients, within an updated planning of the public city and local welfare, which can respond to the growing demand for well-being and urban equality, in quantitative and performance terms. The potential of this approach intersects with the need to make public decision-makers and citizens increasingly aware of the role played by ecosystem services (Saarikoski et al. 2018; Grêt-Regamey et al. 2017) orienting urban planning strategies (Hansen et al. 2015; Cortinovis and Geneletti 2018) towards a dimension of sustainability and resilience of the territory to climate change (McPhearson et al. 2014, 2015). In this perspective, the actions to reduce land take and soil sealing, concerning the protection and appreciation of the ecosystem, are more effective and not only about quantitative parameters but introducing assessment criteria which consider soil quality and ecosystem functions (Polasky et al. 2011). To achieve an ecosystem dimensioning of the urban plan for different land-use transformation scenarios (Geneletti 2013), the knowledge of soil quality is essential to define the land-use planning choices and identify adequate mitigation or compensatory actions and finally to exclude the soil transformability when the ecosystem values cannot be restored. Compensatory measures based upon quantitative criteria (following the principle for which the same amount of urbanised land must be re-naturalised) appear inadequate or even counterproductive where the aim is not to have, indifferently, new green spaces, but maintain ecosystems and related benefits and restore the degraded one.

The integration between the mapping and evaluation of ecosystem services, and the definition of urban planning strategies and decisions, require a truly "transdisciplinary" approach (Costanza 2008). This approach can recompose the fragmentation of the sector-based analytical contributions and overcome the traditional subordination of specialist scientific disciplines (environmental, ecological, agronomic, geological and pedological) in the spatial planning process. In this co-design model, the different areas of expertise cooperate to define spatial planning and development strategies, objectives and actions by verifying and assessing its impacts and benefits on soil functions and values using adequate criteria and indicators. The Strategic Environmental Assessment (SEA) can acquire a new role becoming a tool that, by mapping and evaluating ecosystem functions and services (Geneletti 2016; Ronchi et al. 2020), can guide and monitor planning strategies, affect its objectives, actions and tools, define regulatory and performance devices, which oriented public and private interventions to pursue common interest and well-being.

Recently, the methods and experiments to assess ecosystem services are increasingly accurate and refined (Haase et al. 2014a; Burkhard et al. 2013; Crossman et al. 2013). The more complex challenge is to raise awareness among public decision-makers and citizens of the approach's potential (Gret-Regamey et al. 2017; Costanza et al. 2017), to achieve effective integration between methodologies of classification and evaluation of ecosystem services, and spatial and landscape planning models (Albert et al. 2016; Gómez-Baggethun and Barton, 2013; Grêt-Regamey et al. 2017). The potential of ecosystem analysis in a transdisciplinary integration process with spatial planning is significant at the different planning scales, where the ex ante assessment capacity of the ecosystem functionality becomes an opportunity for protection and conservation and their enhancement.

There is still the risk that studies and research, which provide detailed and articulated classifications and functionalities of ecosystem services, remain limited to the scientific field of research without influencing experiments of new sustainable planning models resilient to climate change. Evaluating ecosystem services may appear to be a low-priority activity with substantial rhetorical value in evoking ecological innovation applied to spatial planning, but in reality unable to effect planning or influence land-use regulations (Cortinovis and Geneletti 2018).

Internationally, urban and landscape planning experiences are becoming more frequent, at the local and regional scales, where there is an explicit relationship between assessment of ecosystem services and definition of land uses (Hansen et al. 2015). These are practices where methodologies have been trialled during decision-making processes (Saarikoski et al. 2018; Cortinovis and Geneletti 2018; Ronchi et al. 2020) to define strategies of urban and landscape planning objectives (Mascarenhas et al. 2014; Haase et al. 2014b), aimed at increasing the multiple benefits provided by ecosystem services, in terms of air quality, water drainage and run-off mitigation, microclimate regulation and pollution reduction (Gómez-Baggethun and Barton 2013; Rall et al. 2015).

In Italy, the experiences assessing ecosystem services integrated directly into the spatial planning process (and Strategic environmental assessment), conditioning its decisions and directly affecting forecasts of soil transformability, are still partial, even if awareness of this approach is increasing (Geneletti et al. 2020). This volume collects some of the most significant experiences in Italy. One limit may be due to the difficulty of communicating to citizens and decision-makers the importance of ecosystem services for the quality of human life (Porter and Kramer 2011) and urban well-being, and the direct relationships with the decisions on land use, which rarely leads them to be considered in policy and planning decisions (Costanza et al. 1997, 2017). It may be helpful to quantify these services in the "market" terms (TEEB 2008, Gómez-Baggethun et al. 2010). The role they play is so essential for human well-being that it is difficult to attribute an economic value (de Groot et al. 2002), but even though there are risks associated with the "monetisation of nature" (Costanza 2006; Gómez-Baggethun and Ruiz-Pérez 2011), research on the economic valuation of ecosystem services has contributed to making the value they have for global and local economies more intelligible. These can be used "freely" by citizens and economic stakeholders (Costanza et al. 1997).

1.4 Green and Blue Infrastructure and Nature-Based Solutions (NBS) for the Resilient Regeneration of the Contemporary City

Recently, Green and Blue Infrastructure (GBI) (European Commission 2013a, b) has played an increasingly significant role in practices and processes of urban and landscape planning (Benedict and McMahon 2000; EEA 2014; Lafortezza et al. 2013) by redefining spatial planning paradigms in a resilient and ecologically oriented way. They have made a planning contribution in the usable ecological reconfiguration of the contemporary city and region. In the document, "An Action Plan for nature, people and economy" (European Commission 2017), the European Commission identified Green Infrastructure as the best management and protection tool for European natural capital sites as priorities to rescue threatened habitats and species in Europe, while pursuing an objective of restoring at least 15% of the degraded ecosystems and maintaining the ecosystems and their services. GBI may not appear a new solution in urban planning and might be seen as "old wine in new bottles" (Davies et al. 2006, Von Christian et al. 2012), if considered design of networks of open spaces with ecological connotations. They are a relevant and fruitful field of experimentation in the re-urbanisation of the contemporary city in a resilient and adaptive dimension (Ahern 2007), which responds to multiple differentiated functions referring to the soil characters and the design scale of the project.

While restoring the methodological and planning tradition of Ecological networks which guarantee biodiversity and connections between highly natural areas (Bennet and Mulongoy 2006), GBI supersedes and re-orientates the concept of network, in the multifunctional and multi-scale perspective (Arcidiacono et al. 2018b). "GBI is a strategically planned network of natural and semi-natural areas with other environmental features designed and managed to deliver a wide range of ecosystem services (…). In addition to providing a key tool to halt and reverse the loss of biodiversity, this network of green and blue spaces provides simultaneously a multiplicity of benefits in a cost-efficient way. The delivery of those benefits is maximised if planned at a strategic level" (European Commission 2013b). GBI is mooted as a planning tool for the conservation and protection of rural and natural systems' landscape values, supporting land take restriction measures while becoming the spatial design of reference for the implementation and consistency check of nature-based solutions (NBS), or "living solutions inspired and supported by the use of natural processes and structures which are designed to address various environmental challenges in an efficient and adaptable manner, while simultaneously providing economic, social, and environmental benefits" (European Commission 2015; Maes and Jacobs 2017). GBI is a tool to increase the quantity and quality of natural resources within the city's central and peri-urban fabrics (European Commission 2013b), where regulating ecosystem services are precious and fragile, from within a project that integrates systems of natural areas and water resources with slow mobility networks, energy and digital infrastructures, building systems of spatial, social and value relationships, cohesive and inclusive, supporting widespread regeneration processes of existing fabrics. Networks of spaces, waters, landscape, urban and agricultural, green areas and places of waste and abandonment interact and penetrate the building fabrics to the core and bring a different contribution to the urban metabolism. This contribution is based upon recycling and optimising resources and social re-appropriation of shared assets. Systems of areas are managed by multilevel governance processes, in which urban planning intersects with spontaneous planning actions, as the "tactical urbanism" (Lyndon and Garcia 2015), which can construct and reinforce local community identity (European Commission 2015).

Designing GBI, open areas and spaces become planning places and components, which identify and differentiate NBS based on ecosystem values, within an overall frame based on the re-composition and ecological and social regeneration of the urban structure. A network of areas, in the urban and peri-urban area, denoted by a specific landscape dimension and ecosystem condition which provide support for city's naturalness reconstruction strategies and impact regulation on soil biological cycles, verifying its permeability and porosity ratios (European Commission 2013b; Maes et al. 2014).

GBI is a supporting structure, which is designed based upon the mapping and evaluation of functionalities and ecosystem services which allow to assess and verify planning decisions and strategies. It is a planning perspective which overturns the traditional quantitative and functional urban planning approach. It introduces a performance dimension at setting performative criteria and design guidelines, to be adapted to physical and social local contexts, to which anchoring the methods of evaluation and control of urban regeneration processes and restriction of land take.

GBI becomes a strategic and spatial "backbone" in public–private negotiation processes, by guiding transformation and regeneration proposals towards updated objectives of collective and public interest – no longer just increasing the supply of areas and services but achieving new ecological performance, which can conserve and enhance the precious and threatened regulating ecosystem services within the city's built-up fabrics.

GBI's multi-scale dimension constitutes its primary potential in territory's spatial structuring. At the landscape scale, it provides a priority structure for planning strategies, which protect the landscape's ecological and natural values, and develops territorial use systems that resiliently and adaptively support guidelines and devices to protect and sustain landscape development and provide the re-composition and regeneration of degraded areas. At the urban scale, it provides a framework for defining and selecting NBS, for the updated and informed design of the public city and local welfare, appropriate to the evolution of social demand and the concept of urban living quality, which is capable of responding to the recent requirements of collective interest and urban well-being based on a resilient and adaptive configuration of urban and spatial structures.

In this sense, GBI faces the reduction of land take through a planning and design perspective by:

- Coordinating the regulatory restrictions of land-use transformations within a landscape networks design

- Protecting and enhancing the ecological porosity of the consolidated urban fabrics
- Conditioning and articulating the implementation of a general strategy aimed at the densification of the existing city using solutions that do not compromise the supply and quality of existing urban open spaces (even private)
- Addressing the regeneration of abandonment and disused places with active measures towards bringing new naturalness and permeability of the soil inside the built-up city and responding resiliently to climate change pressures

1.5 Conclusions

Nowadays, it is imperative to innovate spatial planning paradigm so that it can address new challenges in the design of the contemporary city, in sustainable and resilience perspective of land-use transformation to climate changes. A model must be found that reduces the use of natural resources, primarily soil, through awareness of the values of services that the ecosystem provides and the multiple benefits that they produce for human well-being. The ecosystem approach constitutes an important model, but there are still several critical issues.

While the ES debate is widespread in the academic discourse, in the literature and scientific research, there is a lack of awareness by public decision-makers on the importance of ecosystem services for urban quality and collective well-being. This need for greater awareness is related to urban planning goals towards redefining and expanding the collective interest concerns, where ecosystem services directly affect the quality of life and well-being of citizens.

The second element of weakness concerns the limited integration of the ecosystem approach into spatial planning; until today, the experiences are limited and sector-based. It seems necessary to assume a transdisciplinary perspective in redefining the urban planning paradigm, in which the different technical disciplines co-participate in setting planning decisions. Planners require greater awareness of the contribution that the ecosystem assessment can provide to the redefinition of the spatial plan for re-urbanising the contemporary city in a flexible and ecologically oriented manner which can respond to the challenges imposed by climate change. In defining spatial and landscape planning strategies that favour the regeneration of the existing city over new land take of agricultural and natural soils, it is essential that the "porosity" of urban fabrics is increased, to conserve open spaces and existing greenery and improving the urban natural capital. The densification of the built-up city becomes an agreeable objective only if practised through regeneration (environmental and social) and the partial re-naturalisation of already urbanised and soil sealed areas, where the transformation of land use must contribute to the pursuit of new ecological and environmental goals. Many recent urban redevelopment projects, implemented in European and Italian cities, have achieved results of renewal and real estate development of areas using greenwashing strategies but have been ineffective in increasing ecosystem quality and social cohesion of open spaces – despite being frequently supported by rhetoric of the ecological city and environmental sustainability.

A third critical "knot to untangle" concerns the scale of ES in terms of the relation between ES mapping and evaluating and spatial planning strategies. Mismatches between the scales at which ES are delivered, demanded and governed are recognised as being one of the most important causes of failures in natural resource management and a critical issue in ES adoption for spatial planning. Investigations on "scale definition" insist that ES assessment must consider the ecological processes that ensure the provision of goods and services, and the relevant application level which is central to any ES evaluation and analysis of environmental changes.

Addressing mismatches requires an adjustment between ES ecological processes at the management and planning scale. The adoption of a multi-scale approach could help overcome or at least reduce this critical issue. Mapping and scale issues must be investigated together for ES implementation in planning and assessment processes, as ES evaluation and mapping are often inadequate or ignored during the decision-making process.

The ES assessment mismatch can be solved (or a suitable compromise found) in the adoption of the landscape scale as a logical setting. This is due to the mixture of historical, social, cultural and environmental aspects and dynamics. A landscape approach goes beyond administrative boundaries to focus on conserving the similarity of the landscape structure. Landscape metrics can help assess the benefiting areas which rely on provisioning areas for the delivery of services.

Another key topic is the importance of including ES consideration in the strategic environmental assessment (SEA), providing a window of opportunity to mainstream ES into decision-making processes and planning formally, and the adoption of an ecological compensation method to redefine and improve proposals for land-uses changes.

A relevant contribution which facilitates experimentation of evaluation and mapping of ecosystem services integration and the construction of a new planning model is related to the development of GBI, as an innovative structure for contemporary planning.

The potential of GBI methodology to provide an innovative approach is becoming increasingly important for planning, where these networks can shape the new framework of the contemporary urban and territorial structure, and systems of open spaces (public and private), urban and peri-urban areas, agricultural and natural soil, are integrated as pieces of an ecologically oriented and socially inclusive recreative and environmental project. GBI permits the experimentation of an across-scale approach to the project, in which the wide-scale design of green networks is downscaled at intercommunal and local levels and can activate forms of governance and social sharing of the local project, within a common perspective of improving the quality of living conditions.

This volume aims to provide a scientific and methodological contribution to the trialling of an innovative method in Italian spatial and landscape planning, through a critical reflection on the opportunities and potential connected to the application of ecosystem services and green and blue infrastructures to spatial planning, demonstrating innovative national case study while highlighting critical issues that need to be resolved.

References

Ahern J (2007) Green infrastructure for cities. The spatial dimension. In: Novotny V, Brown P (eds) Cities of the future: towards integrated sustainable water and landscape management. IWA Publications, London, pp 267–283

Albert C, Galler C, Hermes J, Neuendorf F, Haaren C, Von LA (2016) Applying ecosystem services indicators in landscape planning and management: the ES-in-planning framework. Ecol Indic 61:100–113. https://doi.org/10.1016/j.ecolind.2015.03.029

Arcidiacono A (2015) Tra il dire e il fare. Consumo di suolo, leggi e piani. In: Moccia FD, Sepe M (eds) Una politica per le città italiane. INU Edizioni, Roma

Arcidiacono A, Giaimo C., Talia M (2018a) Priorità per il progetto urbanistico. Reti ambientali, nuovi standard e rigenerazione urbana. In: Moccia F D, Sepe M (eds), Sviluppare, rigenerare, ricostruire città. Questioni e sfide contemporanee, INU Edizioni, Roma

Arcidiacono A, Salata S, Ronchi S (2018b) Un approccio ecosistemico al progetto delle infrastrutture verdi nella pianificazione paesaggistica. Urbanistica 159:102–114

Benedict MA, McMahon ET (2000) Green infrastructure. Recreation 37:4–7. https://doi.org/10.4135/9781412973816.n70

Bennett G, Mulongoy KJ (2006) Review of experience with ecological networks, corridors and buffer zones. Secretariat of the Convention on Biological Diversity, Technical Series, 23. https://www.cbd.int/doc/publications/cbd-ts-23.pdf

Brueckner J K (2001) Urban sprawl: lessons from urban economics. In: Gale WG, Pack JR (eds) Brookings-Wharton papers on urban affairs. Brookings Institution, Washington DC

Burkhard B, Kroll F, Nedkov S, Müller F (2012) Mapping ecosystem service supply, demand and budgets. Ecol Indic 21:17–29. https://doi.org/10.1016/j.ecolind.2011.06.019

Burkhard B, Crossman N, Nedkov S et al (2013) Mapping and modelling ecosystem services for science, policy and practice. Ecosyst Serv 4:1–3. https://doi.org/10.1016/j.ecoser.2013.04.005

Cortinovis C, Geneletti D (2018) Ecosystem services in urban plans: what is there, and what is still needed for better decisions. Land Use Policy 70:298–312. https://doi.org/10.1016/j.landusepol.2017.10.017

Cortinovis C, Geneletti D (2019) A framework to explore the effects of urban planning decisions on regulating ecosystem services in cities. Ecosyst Serv 38:100946. https://doi.org/10.1016/j.ecoser.2019.100946

Costanza R (2006) Nature: ecosystems without commodifying them. Nature 443:749. https://doi.org/10.1038/443749b

Costanza R (2008) Ecosystem services: multiple classification systems are needed. Biol Conserv 141:350–352. https://doi.org/10.1016/j.biocon.2007.12.020

Costanza R, d'Arge R, de Groot R, Farber S, Grasso M, Hannon B (1997) The value of the world's ecosystem services and natural capital. Nature 387:253–260. https://doi.org/10.1038/387253a0

Costanza R, de Groot R, Braat L et al (2017) Twenty years of ecosystem services: how far have we come and how far do we still need to go? Ecosyst Serv 28:1–16. https://doi.org/10.1016/j.ecoser.2017.09.008

Crossman ND, Burkhard B, Nedkov S et al (2013) A blueprint for mapping and modelling ecosystem services. Ecosyst Serv 4:4–14. https://doi.org/10.1016/j.ecoser.2013.02.001

Crutzen PJ (1998) Changing atmospheric chemistry: causes and consequences for environment and climate. Eur Rev 6:7–23. https://doi.org/10.1017/S1062798700002969

Crutzen P (2005) Benvenuti nell'Antropocene. L'uomo ha cambiato il clima, la Terra entra in una nuova era. Mondadori, Milano

Davies C, MacFarlane R, McGloin C, Roe M (2006). Green infrastructure planning guide, p. 45. [online] URL: http://www.scribd.com/doc/
de Groot R, Wilson MA, Boumans RMJ (2002) A typology for the classification, description and valuation of ecosystem functions, goods and services. Ecol Econ 41:393–408. https://doi.org/10.1016/S0921-8009(02)00089-7
Dodge R, Daly A, Huyton J, Sanders L (2012) The challenge of defining wellbeing. Int J Wellbeing 2:222–235. https://doi.org/10.5502/ijw.v2i3.4
Douglas I, James P (2015) Urban ecology. An introduction. Routledge, New York. ISBN 9780415538954
EEA (European Environment Agency) (2006) Urban sprawl in Europe – the ignored challenge. Report no 10. Publications Office of the European Union, Luxembourg
EEA (European Environment Agency) (2014) Spatial analysis of green infrastructure in Europe. Report no 2. Publications Office of the European Union, Luxembourg
EEA (European Environment Agency) (2016) Extreme temperatures and health, EEA report. Publications Office of the European Union, Luxembourg
EEA (European Environment Agency) (2017) Climate change, impacts and vulnerability in Europe 2016. An indicator-based report, EEA report no 1. Publications Office of the European Union, Luxembourg
European Commission (2002) Towards a thematic strategy for soil protection, COM(2002)
European Commission (2006a) Thematic strategy for soil protection, COM(2006) 231
European Commission (2006b) Proposal for a directive establishing a framework for the protection of soil, COM(2006)
European Commission (2012) Guidelines on best practice to limit, mitigate or compensate soil sealing. Publications Office of the European Union, Luxembourg
European Commission (2013a) Communication from the Commission to the European Parliament, the Council, the European economic and social Committee and the Committee of the regions. Green Infrastructure (GI) – Enhancing Europe's Natural Capital {SWD(2013) 155 final}. Bruxelles
European Commission (2013b) Building a green infrastructure for Europe. Publications Office of the European Union, Luxembourg. https://doi.org/10.2779/54125
European Commission (2015) Towards an EU research and innovation policy agenda for nature-based solutions & re-naturing cities. Publications Office of the European Union, Luxembourg
European Commission (2016) FUTURE BRIEF: no net land take by 2050? Produced for the European Commission DG Environment by the Science Communication Unit, Bristol
European Commission (2017) An Action Plan for nature, people and the economy. COM(2017) 198 final. http://ec.europa.eu/environment/nature/info/pubs/docs/brochures/Action_plan_brochure_en.pdf
Gasparrini A, Guo Y, Hashizume M et al (2015) Mortality risk attributable to high and low ambient temperature: a multicountry observational study. Lancet 386:369–375. https://doi.org/10.1016/S0140-6736(14)62114-0
Geneletti D (2013) Assessing the impact of alternative land-use zoning policies on future ecosystem services. Environ Impact Assess Rev 40:25–35. https://doi.org/10.1016/j.eiar.2012.12.003
Geneletti D (2016) Ecosystem services for strategic environmental assessment: concepts and examples. In: Geneletti D (ed) Handbook on biodiversity and ecosystem services in impact assessment. Edward Elgar Publishing, pp 41–61
Geneletti D, Cortinovis C, Zardo L, Adem Esmail B (2020) Planning for ecosystem Services in Cities. Springer Nature Switzerland AG
Giaimo C (ed) (2019) Dopo 50 anni di standard urbanistici in Italia. Verso percorsi di riforma. INU Edizioni, Roma. ISBN:9788876031908
Glæsner N, Helming K, de Vries W (2014) Do current European policies prevent soil threats and support soil functions? Sustainability 6:9538–9563. https://doi.org/10.3390/su6129538
Gómez-Baggethun E, Barton DN (2013) Classifying and valuing ecosystem services for urban planning. Ecol Econ 86:235–245. https://doi.org/10.1016/j.ecolecon.2012.08.019

Gómez-Baggethun E, Ruiz-Pèrez M (2011) Economic valuation and the commodification of ecosystem services. Prog Phys Geogr Earth Environ 35(5):613–628., SageJournals. https://doi.org/10.1177/0309133311421708

Gómez-Baggethun E, de Groot R, Lomas PL, Montes C (2010) The history of ecosystem services in economic theory and practice: from early notions to markets and payment schemes. Ecol Econ 69:1209–1218. https://doi.org/10.1016/j.ecolecon.2009.11.007

Grêt-Regamey A, Sirén E, Brunner SH, Weibel B (2017) Review of decision support tools to operationalize the ecosystem services concept. Ecosyst Serv 26:306–315. https://doi.org/10.1016/j.ecoser.2016.10.012

Haase D, Larondelle N, Andersson E, Artmann M et al (2014a) A quantitative review of urban ecosystem service assessments: concepts, models, and implementation. Ambio 43:413–433. https://doi.org/10.1007/s13280-014-0504-0

Haase D, Frantzeskaki N, Elmqvist T (2014b) Ecosystem services in urban landscapes: practical applications and governance implications. Ambio 43(4):407–412. https://doi.org/10.1007/s13280-014-0503-1

Haines-Young R., Potschin M. (2013) Common international classification of ecosystem services (CICES, version 4.3), report to the European Environment Agency, September, pp 1–17

Hansen R, Frantzeskaki N, McPhearson T, Rall E et al (2015) The uptake of the ecosystem services concept in planning discourses of European and American cities. Ecosyst Serv 12:228–246. https://doi.org/10.1016/j.ecoser.2014.11.013

ISPRA (2016) Consumo di suolo, dinamiche territoriali e servizi ecosistemici, Rapporto 2016, Roma

ISPRA (2018) Consumo di suolo, dinamiche territoriali e servizi ecosistemici, Rapporto 2018, Roma

ISPRA (2019) Consumo di suolo, dinamiche territoriali e servizi ecosistemici, Rapporto 2019, Roma

Lafortezza R, Davies C, Sanesi G, Konijnendijk CCC (2013) Green infrastructure as a tool to support spatial planning in European urban regions. iForest – Biogeosci For 6:102–108. https://doi.org/10.3832/ifor0723-006

Lefevre H (1968) Le droit à la Ville. Anthropos, Paris

Lyndon M, Garcia A (eds) (2015) Tactical urbanism. Short-term action for long-term change. Island Press, Washington, DC https://doi.org/10.5822/978-1-61091-567-0

Maes J, Jacobs S (2017) Nature-based solutions for Europe's sustainable development. Conserv Lett 10:121–124. https://doi.org/10.1111/conl.12216

Maes J, Barbosa A, Baranzelli C et al (2014) More green infrastructure is required to maintain ecosystem services under current trends in land-use change in Europe. Landsc Ecol 30:517–534. https://doi.org/10.1007/s10980-014-0083-2

Maes J, Zulian G, Thijssen M, Castell C, Baró F, Ferreira AM, Melo J (2016) Mapping and assessment of ecosystems and their services: urban ecosystems. Publications office of the European Union, Luxembourg

Mascarenhas A, Ramos TB, Haase D et al (2014) Integration of ecosystem services in spatial planning: a survey on regional planner's views. Landsc Ecol 29:1287–1300. https://doi.org/10.1007/s10980-014-0012-4

McPhearson T, Hamstead ZA, Kremer P (2014) Urban ecosystem services for resilience planning and management in New York City. Ambio 43(4):502–515. https://doi.org/10.1007/s13280-014-0509-8

McPhearson T, Andersson E, Elmqvist T, et al (2015) Resilience of and through urban ecosystem services. Ecosyst Serv 12:152–156, Elsevier. https://doi.org/10.1016/j.ecoser.2014.07.012

Millenium Ecosystem Assessment (2005) Ecosystems and human. Well-being: our human planet. Island Press, Washington, DC. ISBN:9781559633864

Polasky S, Nelson E, Camm J et al (2008) Where to put things? Spatial land management to sustain biodiversity and economic returns. Biol Conserv 141:1505–1524. https://doi.org/10.1016/j.biocon.2008.03.022

Polasky S, Nelson E, Pennington D, Johnson KA (2011) The impact of land-use change on ecosystem services, biodiversity and returns to landowners: a case study in the state of Minnesota, pp 219–242. https://doi.org/10.1007/s10640-010-9407-0

Porter ME, Kramer MR (2011) Creating shared value. Harward Business Rev 89(1–2):62–77

Rall EL, Kabisch N, Hansen R (2015) A comparative exploration of uptake and potential application of ecosystem services in urban planning. Ecosyst Serv 16:230–242. https://doi.org/10.1016/j.ecoser.2015.10.005

Ronchi S, Salata S, Arcidiacono A et al (2019) Policy instruments for soil protection among the EU member states: a comparative analysis. Land Use Policy 82:763–780. https://doi.org/10.1016/j.landusepol.2019.01.017

Ronchi S, Arcidiacono A, Pogliani L (2020) Integrating green infrastructure into spatial planning regulations to improve the performance of urban ecosystems. Insights from an Italian case study. Sustain Cities Soc 53:1–12. https://doi.org/10.1016/j.scs.2019.101907

Saarikoski H, Primmer E, Saarela SR et al (2018) Institutional challenges in putting ecosystem service knowledge in practice. Ecosyst Serv 29:579–598. https://doi.org/10.1016/j.ecoser.2017.07.019

Schewenius M, McPhearson T, Elmqvist T (2014) Opportunities for increasing resilience and sustainability of urban social–ecological systems: insights from the URBES and the cities and biodiversity outlook projects. Ambio 43(4):434–444. https://doi.org/10.1007/s13280-014-0505-z

Shekhar H, Schmidt AJ, Wehling HW (2019) Exploring wellbeing in human settlements – a spatial planning perspective. Habitat Int 87:66–74. https://doi.org/10.1016/j.habitatint.2019.04.007

TEEB (2008) The economics of ecosystems and biodiversity: an interim report. European Commission, Brussels

UN (United Nations) (2019a) Global sustainable development report 2019: the future is now – science for achieving sustainable development. United Nations, New York

UN (United Nations) (2019b) World urbanization prospects. The 2018 revision. United Nations, New York

Von Christian A, Von Haaren C, Galler C (2012) Ecosystem services. Old wine in new bottles or an incentive for German landscape planning? Naturschutz und Landschaftsplanung 44(5):142–148

WHO (World Health Organization) (2019) Health, environment & sustainable development goals. New York

Chapter 2
Urban Green Infrastructure: Opportunities and Challenges at the European Scale

Grazia Zulian, Julie Raynal, Rayka Hauser, and Joachim Maes

Abstract Investing in the deployment of Green Infrastructure (GI) can provide many different environmental, societal and economic benefits. High-quality, biodiversity-rich blue-green areas increase land sustainability and help to solve many challenges, such as air pollution, noise, climate change impacts, heat waves, floods and public health concerns. In order to design and manage an efficient GI, there is a need for consistent multi-scale and inter-sectorial GI-related policies. During the last six years, the Commission developed and implemented a set of initiatives to sustain the deployment of multi-scale GI. This chapter provides an overview of the most recent documents and initiatives released in the framework of EU biodiversity strategy to 2020, the EU Strategy on Green Infrastructure and the Action plan for nature, people and the economy.

Keywords Science for policy · Green infrastructure · Land sustainability · Ecosystem services · Multi-functionality · Ecosystem restoration · Healthy ecosystems · Urban ecosystems

G. Zulian (✉) · J. Maes
European Commission Directorate-General, Joint Research Centre, Directorate D Sustainable Resources, Unit D3 – Land Resources, Ispra, Italy
e-mail: grazia.zulian@ec.europa.eu

J. Raynal
European Union, European Commission, Directorate-General for Environment, Quality of Life, Industrial Emissions and Safety (ENV.C.4), Bruxelles/Brussel, Belgium

R. Hauser
European Union, European Commission, Directorate-General for Environment, Natural Capital, Biodiversity (ENV.D.2), Bruxelles/Brussel, Belgium

A. Arcidiacono, S. Ronchi (eds.), *Ecosystem Services and Green Infrastructure*, Cities and Nature, https://doi.org/10.1007/978-3-030-54345-7_2

2.1 Introduction

Green infrastructure is defined in the EU green infrastructure strategy as 'a strategically planned network of natural and semi-natural areas with other environmental features designed and managed to deliver a wide range of ecosystem services. It incorporates green spaces (or blue if aquatic ecosystems are concerned) and other physical features in terrestrial (including coastal) and marine areas. On land, green infrastructure is present in rural and urban settings' (EC 2013). Urban Green Infrastructure (UGI) refers to the strategically managed network of urban green spaces and natural and semi-natural ecosystems situated within the boundary of an urban ecosystem. In a highly urbanised landscape, it represents a key element for the deployment of an integrated and resilient GI.

As a natural, service-providing infrastructure, it is often more cost-effective, more resilient and more capable of meeting social, environmental and economic objectives than 'grey' infrastructure. GI can help addressing the challenges of the twenty-first century and its use should be further developed.

Green Infrastructures are multi-functional and have the capacity to provide valuable support to multiple and integrated policies. High-quality and biodiversity-rich areas contribute solutions to many challenges, such as air pollution, noise, climate change impacts, heat waves, floods and public health concerns.

In the first section of the chapter, we will provide a synthesis of the most recent documents and initiatives promoted by the European Commission to support the deployment of an effective and integrated GI at a continental scale. The second section provides an EU scale application of mapping Urban GI and Urban Ecosystem services in European cities.

2.2 EU Initiatives for the Deployment of Multi-scale Green Infrastructure

The deployment of a resilient, integrated and multi-scale GI is the backbone for a new framework for policies aimed at preserving Europe's natural environment.

> *Climate change, biodiversity, food security, deforestation and land degradation go together. We need to change the way we produce, consume and trade. Preserving and restoring our ecosystem needs to guide all of our work. We must set new standards for biodiversity cutting across trade, industry, agriculture and economic policy* (Von Der Leyen 2019).

Urban and peri-urban nature, green areas and forest become an important part of the solution if embedded in a multi-scale approach together with restoration of degraded ecosystems, conservation and nature protection.

> *Climate and environmental degradation are global challenges with a very local dimension. Local and regional authorities have a key role in addressing them, including through the strategic deployment of green and blue infrastructure, and the integration of ecosystems and their services into decisions. Enabling factors include political will, stakeholder*

engagement and support, know-ledge generation and use, practical tools for planners and practitioners as well as adequate resources. (Take away message, *Session on: EU level Green Infrastructure and Ecosystem Services in the decision-making process*) (EU week of Regions and Cities 7–10 October 2019)

Recently new documents[1] and initiatives have been released in the framework of, the EU biodiversity strategy to 2020, the EU Strategy on Green Infrastructure and the Action plan for nature, people and the economy. The Action Plan aims to improve the practical implementation of the EU nature legislation and accelerate progress towards the EU 2020 goal of halting and reversing the loss of biodiversity and ecosystem services:

- Review of progress on implementation of the EU green infrastructure strategy (EC 2019f)
- Additional information on the review of implementation of the EU green infrastructure strategy (EC 2019a)
- EU guidance on supporting strategic EU-level green and blue infrastructure projects (EC 2019e)
- EU guidance on integrating ecosystems and their services into decision-making (EC 2019b, c, d)

The first two documents (EC 2019a, f) address the progress made and challenges encountered at both EU and Member State (MS) level in carrying out the GI Strategy. They draw some lessons and put forward suggestions for the further implementation of the strategy. The documents provide relevant examples of sectorial GI-related policies implemented at various governance levels. Some examples are:

- Agricultural Policy, Rural Development Program (RDP) – 'restoring, preserving and enhancing ecosystems related to agriculture and forestry'.
- Forestry – Forest and Urban forests are one of the building blocks of an integrated and multi-level GI.
- Spatial Planning and Urban Policy – GI-related objectives have been integrated into the spatial planning policies in several MS with requirements of ecological connectivity, prevention of fragmentation, the establishment of national ecological networks.
- Transport Policy.
- Water Management Policy.
- Climate change and disaster risk policies.
- Maine and Coastal Policy.
- Leisure and Tourism Policy.

The EU urban policy is providing an interesting contribution to the GI deployment.

[1] https://ec.europa.eu/environment/nature/ecosystems/index_en.htm

GI has been included in the award criteria of the European Green Capital[2] and Green Leaf awards.[3]

Under the Urban Agenda for the EU,[4] a Partnership on Sustainable use of land and nature-based solutions was launched in 2017.[5] Action 9 of the endorsed Action Plan aims at 'Agreeing on common targets and indicators for NBS, urban green infrastructure, biodiversity and ecosystem services in cities'.

Knowledge of urban GI is also improving with the support of the MAES 'EnRoute'[6] project, and Horizon 2020 projects on nature-based solutions implemented in urban areas[7] (Somarakis, Stagakis and Chrysoulakis, 2019). Several initiatives launched by European cities target GI at city and local levels as well.[8]

The EU guidance on supporting strategic EU-level green and blue infrastructure projects[9] aims to support a strategic, multi-scale and integrated approach to the deployment of green and blue infrastructure in order to maximise the delivery of ES and connected benefits. It also provides information on relevant existing funding sources and supporting tools. The aim is to improve the connectivity between Natura 2000 network and other high-value areas important for biodiversity and ES. This guidance complements the dedicated guidance on integrating ecosystems and their services in planning and decision-making also foreseen under the Action Plan. The guidance is complemented by a Joint Science for Policy Report[10] that present complementary GI mapping approaches (Estreguil et al. 2019).

The Commission guidance on integrating ecosystems and their services into decision-making helps to better taking into account the environmental, social and economic benefits provided by GI and urban GI. The guidance consists in three volumes and includes:

- An extensive conceptual framework for the integration of Ecosystem Services into decision-making
- A consistent set of examples of Ecosystems and their services implemented at different Policy levels
- A section providing methodologies, applications and examples for mapping and assessing Ecosystem Services

[2] https://ec.europa.eu/environment/europeangreencapital/

[3] https://ec.europa.eu/environment/europeangreencapital/europeangreenleaf/

[4] http://www.urbanagendaforthe.eu

[5] https://ec.europa.eu/futurium/en/sustainable-land-use; https://ec.europa.eu/futurium/en/system/files/ged/sul-nbs_finalactionplan_2018.pdf

[6] www.oppla.eu/EnRoute

[7] E.g. Nature4Cities, GrowGreen, NAIAD, NATURVATION, UNALAB, Connecting and UrbanGreenUp

[8] https://www.wien.gv.at/stadtentwicklung/studien/pdf/b008440.pdf

[9] The Guidance contributes to implementing Action 12 of the Action Plan for Nature, People and the Economy.

[10] https://ec.europa.eu/jrc/en/publication/strategic-green-infrastructure-and-ecosystem-restoration

The European MAES (Mapping and Assessment of Ecosystems and their Services) initiative[11] has developed a coherent analytical framework to ensure the implementation of consistent approaches linking biodiversity, ecosystem condition and ecosystem services. The framework has been implemented across Member States, at the EU level (first MAES Report, 2013)[12] and has been tested at urban scale in the EnRoute[13] project (Zulian et al. 2018; Maes et al. 2019).

EnRoute stands for Enhancing Resilience of urban ecosystems through green infrastructure. It was a project of the European Commission in the framework of the EU Biodiversity Strategy and the Green Infrastructure Strategy, which aimed at encouraging the exchange of experiences within local municipalities across Europe.

EnRoute was the MAES community of practice on urban ecosystems (cities, researchers, member states, EU services). The main objectives of the project are as follows:

- Operationalise the urban MAES indicator framework – the framework was implemented in 21 city labs and at European scale.
- Analyse how science supports urban policy – the Science Policy Interface was explored using a semi-qualitative approach.
- Enhance networking activity on urban GI – city labs and EU services shared information on OPPLA platform (https://oppla.eu/groups/enroute) and contributed to Action 9 of the Partnership on Sustainable use of land and nature-based solutions.

2.3 Land Configuration and Urban Ecosystem Condition and Services in European Cities[14]

In 2017–2018, in the framework of the EnRoute project, urban ecosystem condition and services were mapped and assessed in 700 European cities (Zulian et al. 2018; Maes et al. 2019). The assessment framework was derived from the fourth and fifth MAES reports (Maes et al. 2016, 2018). Table 2.1 shows the list of indicators.

The study focused on 700 European cities and their surroundings. As basic mapping boundaries and spatial reporting units, the Functional Urban Areas (FUA), or Spatial system for city statistic, was used (Dijkstra and Poelman 2012; Urban Europe 2016; EuroStat 2017). A FUA consists in a[15] core city and a commuting zone, meaning a number of smaller municipalities gravitating around the core city

[11] https://biodiversity.europa.eu/maes

[12] All the key documents published under the MAES initiative can be found in https://biodiversity.europa.eu/maes/#REPORTS

[13] https://oppla.eu/groups/enroute

[14] The figures and the section are part of the Final EnRoute Report (Maes et al. 2019).

[15] Usually a FUA has one core city, but it can happen to have multiple core cities within the same FUA (Eurostat 2017).

Table 2.1 Indicators of urban ecosystem condition and urban ecosystem services for which data were available at European scale (Maes et al. 2019)

Sub-category	Indicator (unit of measure)		Reporting unit
Pressures			
Emissions of air pollutants	Emissions of NOx (tons/year)		FUA
Invasive alien species	Negative impact of terrestrial invasive alien species (dimensionless)		
Environmental quality			
Landscape composition	Land mosaic types (%)		FUA
Population	Population density (inhab./km^2)		Core city
	Relative lived density (inhab./km^2 of artificial land)*		FUA
Soil sealing	Sealed soil per surface (%)		Core city
	Sealed surface (m^2) per inhabitants		
	Inhabitants per sealed surface (m^2) in core cities		
	Sealed soil per land mosaic type(s)* (%)		
Air pollutants	PM $_{10}$ yearly average (μg/m^3)		FUA
	PM $_{10}$ 90.4 percentile of daily mean PM10 concentrations		
	O_3 26th highest daily maximum 8-hour value in μg/m^3		
	NO_2 yearly average (μg/m^3)		
Structural ecosystem attributes			
Green areas in core cities (public and UGI)	Publicly accessible green	m^2/inhabitant	Core city
		m^2 of green/city surface	
	All urban green	m^2/inhabitant	
		m^2 of green/city surface	
		Share of urban green areas with slope up to 25 percent (% over the total urban green areas)	
Spatial configuration of urban green areas	Spatial integrity: m^2 (of GI typology*)/total m^2 GI (%)		Core city
Urban protected areas	Natura 2000/FUA surface (%)		FUA
	Natura 2000/city surface (%)		Core city
	Share of Natura 2000 sites within FUAS/total Natura 2000 (%)		Per country
Ecosystem services	Nature-based recreation		FUA
	Urban pollination		FUA
	Flood control		Core City

(Eurostat, 2017). In the Member states of the European Union (EU-28), the system consists in 700 FUAs, 900 core cities and 600 commuting zones.

The FUA system is especially useful for a multi-scale assessment, when there is a need to focus on cities considering their local ecosystems, the regional interactions between connected cities and the complexity or urban ecosystems at a continental scale.

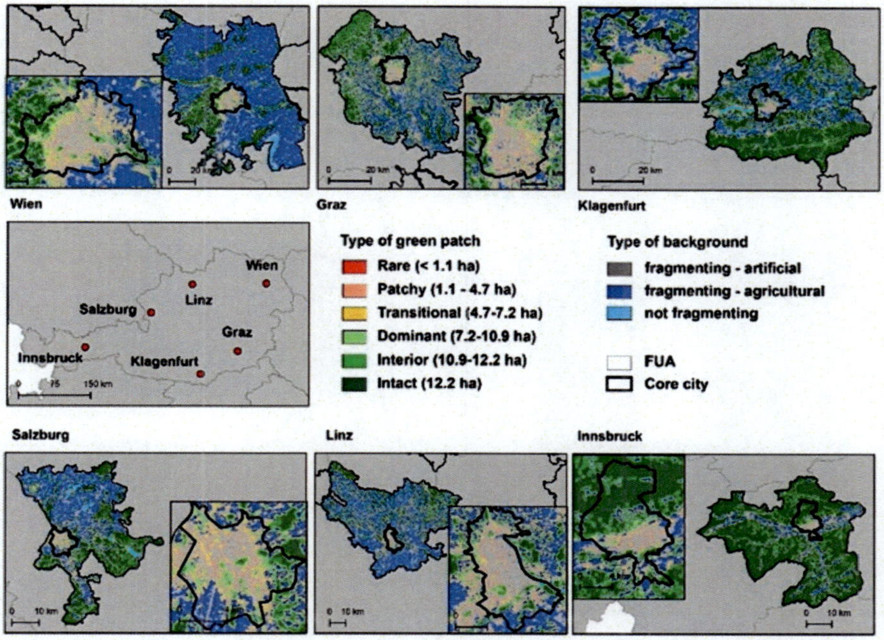

Fig. 2.1 Spatial Integrity indicator mapped in Austrian cities (Source data Urban Atlas [https://land.copernicus.eu/local/urban-atlas] – tool Guidos Tool Box https://forest.jrc.ec.europa.eu/en/activities/lpa/gtb/; Vogt and Riitters 2017)

All indicators presented in Table 2.1 are spatially explicit and maps are available for all FUA systems. See as an example Figs. 2.1 and 2.2, where the structure of Unban GI was mapped within the FUAs (in this case, Austrian cities are presented as an example) and reported at the EU scale.

In order to assess the performance of cities, FUAs have been clustered in 6 types attributable to three fairly homogenous sets of characteristics derived from population density and co-occurrence of dominant land types (namely artificial, natural and semi-natural and agricultural) as shown in Fig. 2.3. Cities are not islands; they represent a connected system of places where people live, work and interact. The way cities are organised and perform is strongly related to their ecological, geographical, economic and socio-political context. Land configuration within a FUA, or of a system of FUAs, strongly affects its performance especially when considering Green Infrastructure and its capacity to provide ES.

The behaviour of the set of indicators of urban ecosystem condition and ecosystem services demonstrated that the typology of cities with respect to urban green infrastructure and their embedment in the surrounding is highly relevant (Table 2.2).

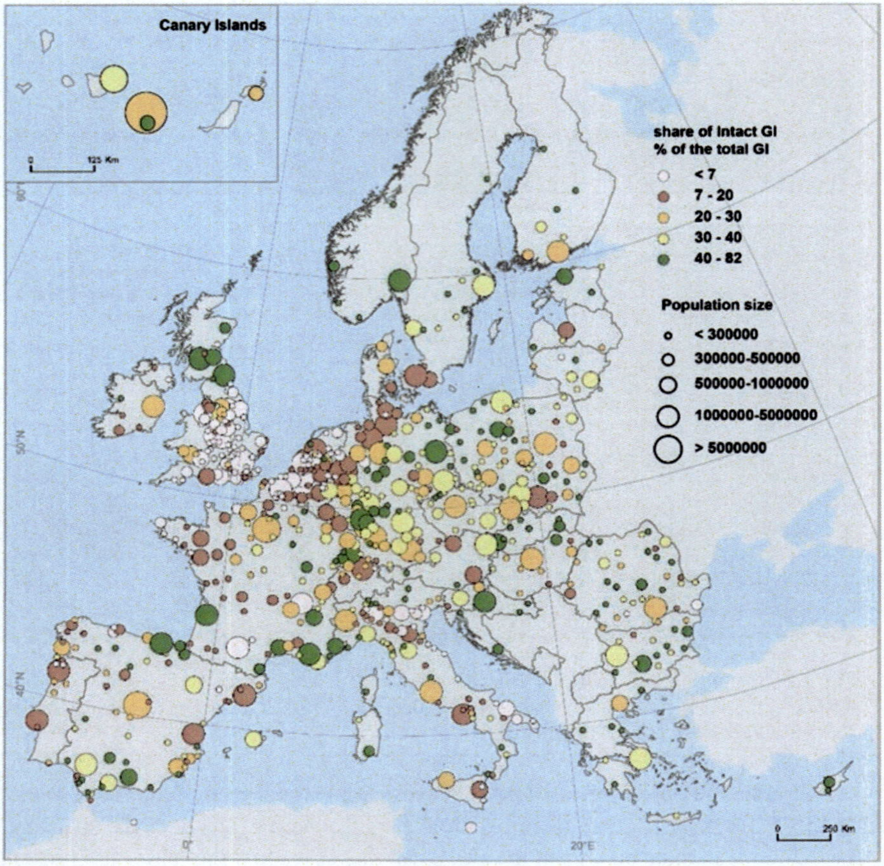

Fig. 2.2 Map of share of intact GI elements within FUAs

2.4 Conclusion

The European GI strategy and the EU biodiversity Strategy to 2020 provided an updated set of resources for the implementation of sectorial GI-related policies and for the integration of Ecosystem Services into decision-making. The MAES initiative demonstrated how Ecosystem condition and services can be mapped to support policies at the European, National and local scales. The Horizon 2020 programme on nature-based solutions supported the convergence of several frameworks for monitoring green infrastructure and nature-based solutions.

The Commission has been providing the tools enabling a wider use and development of green and blue infrastructure in the EU. Now Member States, local and regional governments and civil society have to make use of those tools and

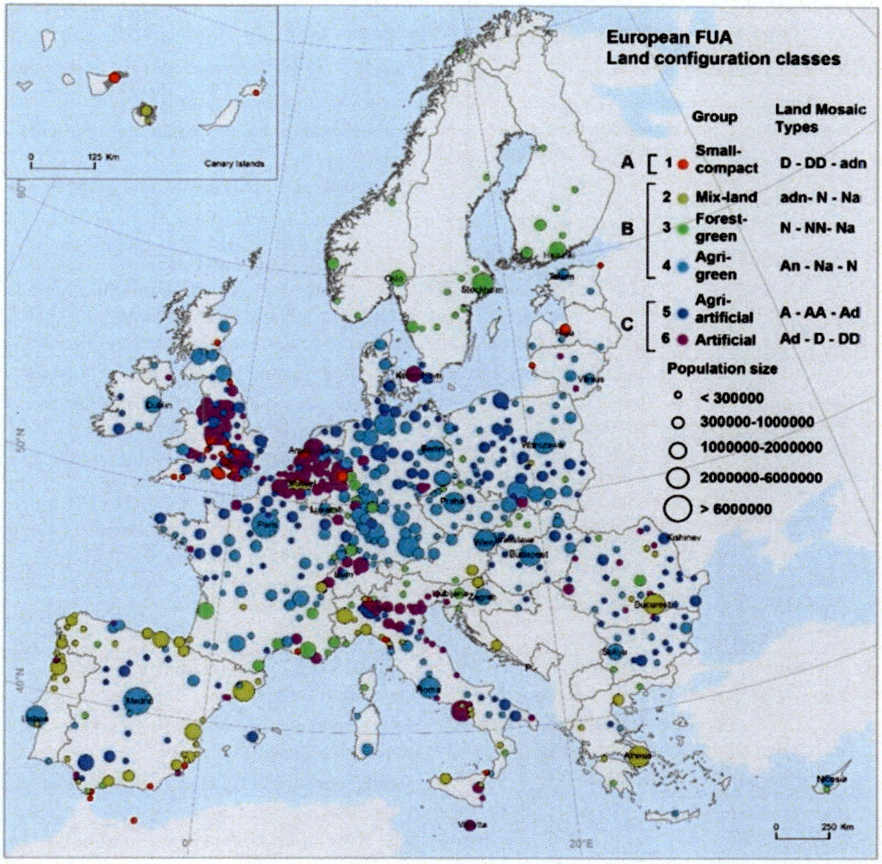

Fig. 2.3 Spatial distribution of European functional urban areas with reference to share of dominant land types and population density. The map includes FUAs in Norway and Switzerland (Maes et al. 2019)

implement ambitious policies and actions so that these tools can actually deliver benefits.

The New Green Deal and the upcoming EU biodiversity strategy to 2030, which will be adopted during the first 100 days of the new European Commission, provide a unique opportunity to deliver an operational indicator framework for measuring the impact of urban green infrastructure and nature-based solutions in cities.

Table 2.2 FUA typologies classified according to share of dominant land types and population density and relative ecosystem condition and ecosystem services (Maes et al. 2019)

Typology	Description	Ecosystem condition and ecosystem services
1 – Small compact FUAs	Characterised by a relatively high population density, often with the absence of commuting zone (73% of the cities belonging to type 1 do not have a respective commuting zone)	Higher average share of soil sealing
		Largest area with a low capacity to mitigate floods
		Highest share of publicly available green space (urban parks), measured as a percentage over the total area of the core city and good score for cultural ecosystem services. But in absolute terms, the average area measured in squared metres per inhabitant and the range around this average is lower than in other city types
2 – Mixed land cover FUAs	Cities characterised by 'mixed land' or high heterogenic anthropic activity and a relatively remarkable difference between population density in core city and surroundings. A mixed land cover corresponds to an absence of a dominant land type	Highest population density of all city types
		No other city type has higher ozone levels. In fact, the distribution of average ozone concentrations over the different city types matches expectations well. Ozone concentrations are typically higher in rural areas relative to urban areas where ozone is degraded following reaction with NO released by traffic
		This city type has for most other indicators average values and ranks mostly between other city types
3 – Forest FUAs	The presence of natural ecosystem types (dominance of peri-urban GI) characterises this group, together with a relatively low population density and the presence of areas of transition to small patches of agriculture	Highest values for regulating ecosystem services and for recreation potential
		Lower than average level of soil sealing and population density
		Score better for air quality indicators than other city types
		The share of urban green space is high (as is the share of Natura 2000), but this seems to go at the cost of the share of public urban green space (urban parks)
4 – Agri-green FUAs	This group is characterised by presence of agriculture and transition to semi-natural areas in a relatively vast surface	Agri-green and agri-artificial cities, which are closer to the European average. The agri-artificial type is characterised by lower air quality than agri-green, at least for PM10 and NOx. Agri-green cities have higher values for urban green space indicators including coverage by the Natura 2000 network and for ecosystem services than agri-artificial. This latter type exhibits poor values on the recreation indicators
5 – Agri-artificial FUAs	This group presents dominance of agriculture and transition to artificial (land mosaic class) in a relatively vast surface with low FUA population density	

(continued)

Table 2.2 (continued)

Typology	Description	Ecosystem condition and ecosystem services
6 – Artificial FUAs	This group shows a prevalence of artificial areas in transition from–to agriculture. Population density is very high. There is no significant difference between population density within and outside core cities, which means that we find high population densities also outside the city	Cities with a predominance of artificial land cover are clearly performing worse than other city types with respect to the two pressure indicators (NOx emissions and invasive alien species), soil sealing, air quality and coverage by Natura 2000. They have a lower share of urban green infrastructure but perform reasonably well in terms of public green infrastructure. They have a low capacity for flood control. Also suitability for pollinators is low in this type

Nevertheless 34% of Forest-green cities are in Nordic Countries (Norway, Finland and Sweden) where people can benefit of the freedom to roam, or 'Everyman's right'. 'Everyman's right' is the general public's right to access certain public or privately owned land, lakes, and rivers for recreation and exercise. The availability of public parks is compensated by the opportunities provided by the Everyman's right

References

Dijkstra and Poelman (2012) Cities in Europe, the new OECD-EC definition, a series of short papers on regional research and indicators produced by the Directorate-General for Regional and Urban Policy. http://ec.europa.eu/regional_policy/sources/docgener/focus/2012_01_city.pdf

EC (2013) Green Infrastructure (GI) — enhancing Europe's natural capital, COM(2013) 249 final, p 13. https://doi.org/10.1017/CBO9781107415324.004

EC (2019a) Additional information on the review of implementation of the green infrastructure strategy, SWD(2019) 184 final. https://doi.org/10.1017/CBO9781107415324.004

EC (2019b) EU guidance on integrating ecosystems and their services into decision-making – part 1/3, SWD(2019) 305 final. Available at: http://ec.europa.eu/environment/nature/ecosystems/index_en.htm

EC (2019c) EU guidance on integrating ecosystems and their services into decision-making – part 2/3, SWD(2019) 305 final. doi: SWD(2013) 527

EC (2019d) EU guidance on integrating ecosystems and their services into decision-making – part 3/3, SWD(2019) 305 final. https://doi.org/10.1017/CBO9781107415324.004

EC (2019e) Guidance on a strategic framework for further supporting the deployment of EU-level green and blue infrastructure. Available at: https://ec.europa.eu/environment/nature/ecosystems/index_en.htm

EC (2019f) Review of progress on implementation of the EU green infrastructure strategy, COM(2019) 236 final. EC, Brussels. https://doi.org/10.1017/CBO9781107415324.004

Estreguil C et al. (2019) Strategic green infrastructure and ecosystem restoration. https://doi.org/10.2760/36800

Eurostat (2017) Methodological manual on city statistics – 2017 edition. Publications Office of the European Union, Luxembourg. https://doi.org/10.2785/765198

Maes J, Zulian G, Thijssen M, Castell C, Baró F, Ferreira AM, Melo J, Garrett CP, David N, Alzetta C, Geneletti D; Cortinovis C, Zwierzchowska I, Louro Alves F, Souto Cruz C, Blasi C, Alós Ortí MM, Attorre F, Azzella MM, Capotorti G, Copiz R, Fusaro L, Manes F, Marando F, Marchetti M, Mollo B, Salvatori E, Zavattero L, Zingari PC, Giarratano MC, Bianchi E, Duprè E, Barton D, Stange E, Perez-Soba M, van Eupen M, Verweij P, de Vries A, Kruse H,

Polce C, Cugny-Seguin M, Erhard M, Nicolau R, Fonseca A, Fritz M, Teller A (2016) Mapping and assessment of ecosystems and their services. urban ecosystems. Publications Office of the European Union, Luxembourg. https://ec.europa.eu/environment/nature/knowledge/ecosystem_assessment/pdf/102.pdf

Maes J, Teller A, Erhard M, Grizzetti B, Barredo JI, Paracchini ML, Condé S, Somma F, Orgiazzi A, Jones A, Zulian A, Petersen JE, Marquardt D, Kovacevic V, Abdul Malak D, Marin AI, Czúcz B, Mauri A, Loffler P, Bastrup-Birk A, Biala K, Christiansen T, Werner B (2018) Mapping and assessment of ecosystems and their services: an analytical framework for ecosystem condition. Publications office of the European Union, Luxembourg.https://ec.europa.eu/environment/nature/knowledge/ecosystem_assessment/pdf/5th%20MAES%20report.pdf

Maes J et al (2019) Enhancing resilience of urban ecosystems through green infrastructure (EnRoute) final report. EC, Luxembourg. https://doi.org/10.2760/602928

Somarakis G, Stagakis S, Chrysoulakis N (eds) (2019). ThinkNature nature-based solutions handbook. ThinkNature project funded by the EU Horizon 2020 research and innovation programme under grant agreement No. 730338. https://doi.org/10.26225/jerv-w202

Urban Europe (2016) Statistics on cities, towns and suburbs (ISBN: 978-92-79-60139-2; https://doi.org/10.2785/91120; Cat. No: KS-01-16-691-EN-N). https://ec.europa.eu/eurostat/statistics-explained/index.php?title=Urban_Europe_%E2%80%94_statistics_on_cities,_towns_and_suburbs

Vogt P, Riitters K (2017) GuidosToolbox: universal digital image object analysis. Eur J Remote Sens. Taylor & Francis 50(1):352–361. https://doi.org/10.1080/22797254.2017.1330650

Von Der Leyen U (2019) A Union that strives for more My Agenda for Europe, political guidelines for the next European Commission 2019–2024, p 24. Available at: https://ec.europa.eu/commission/sites/beta-political/files/political-guidelines-next-commission_en.pdf

Zulian G et al (2018) Enhancing resilience of urban ecosystems through green infrastructure (EnRoute). Progress report. Publications Office of the European Union, Luxembourg. https://doi.org/10.2760/958542

Chapter 3
The Green City: From a Vision to a Concept from National to European Perspectives

Jürgen H. Breuste

Abstract The Green City develops during the last decades from a vision to a goal and to a concept. This starts from the roots by NGOs and encouraged people, from national authorities and communities and from the European level. The discussion in some Central and Northern European countries is actually already further developed and bases on concrete steps forward to targets. A key issue is the urban green infrastructure and its networking characteristic, including all forms of nature and the target to let all urban residents equally participate in the benefits urban green areas provide. This brings the concepts of ecosystem services into the centre of interest and makes it available to become a tool in sustainable urban planning. Exactly this is now necessary and in small steps ongoing in several European examples. The urban nature as urban green part of urban planning becomes more valued and is seen as provider of nature-based solutions in cities. The Green City bases on valued, accepted, accessible and benefits for people providing concept, equal to other urban planning concepts. It starts normally with small steps in neighbourhoods and can expand to districts and become a strategy for whole cities.

Keywords Green City · Urban green-blue infrastructure · Biological diversity · Urban nature · Nature protection · National urban strategies · Nature awareness · New Urban Agenda · Urban green space strategies

---

J. H. Breuste (✉)
Department Geography and Geology, Paris Lodron University of Salzburg, Salzburg, Austria
e-mail: juergen.breuste@sbg.ac.at

A. Arcidiacono, S. Ronchi (eds.), *Ecosystem Services and Green Infrastructure*, Cities and Nature, https://doi.org/10.1007/978-3-030-54345-7_3

3.1 Green City – Vision, Goals and Targets

A "Green City" is a city in which all forms of nature (including living organisms, communities and habitats) are of high significance and are important aspects of urban infrastructure and are maintained and expanded for the services they provide for urban residents. Urban nature is the ideal provider of benefits and the concept for urban planning and development. In this regard, urban nature includes the entirety of all existing nature elements present in urban areas, including the relationship between their usage (by residents) and ecological function.

National organizations and the European association of garden, landscape, and sports ground construction (ELCA) promote the concept of a "Green City" as a city built upon the foundation of urban nature. For several decades, these associations, together with scientists and planners, have been arguing for urban development (policies) that include urban nature as an essential component, without constantly using the term "Green City."

The "Green City" is a visionary objective that obligates individuals, organizations and decision-makers to integrate its conceptual vision into their decisions. However, it is not merely a concrete concept that simply needs to be applied. The concept of a "Green City" requires the continuous adaptation of visions under concrete natural, special and social framework conditions. This is only possible if all involved individuals follow the basic principles of this development model and not only understand that urban nature is essential for any liveable, healthy and (biologically) diverse city, but also implement concrete operational objectives. Thus, the "Green City" is developed from the bottom up as well as from the top down.

The term "ecologically oriented urban development" was already coined in the 1980s, yet remains a superficial but vague concept. In many areas, including Europe, there are typically no national policies regarding urban nature (the use and development of urban nature).

In Germany, principles for the promotion of urban nature in urban planning were already developed in the 1980s as guidelines for the implementation of nature conservation in urban development (German: Leitlinien für die Umsetzung des Naturschutzes in die Stadtplanung, Sukopp and Sukopp 1987). These guidelines for the conservation of urban nature are still relevant and applicable today. Urban nature has always been a nucleus of any "Green City" (Fig. 3.1).

3.1.1 Guidelines for the Implementation of Nature Conservation in Urban Planning (Example Germany)

Principle of

1. Priority areas for nature and environmental protection
2. Locally differentiated focus areas for nature protection and landscape management

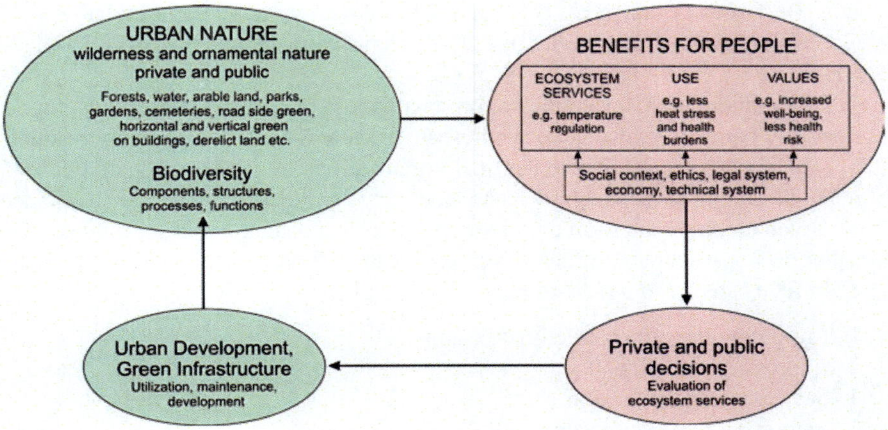

Fig. 3.1 The concept of urban nature (Source: Breuste und Endlicher 2017, changed)

3. Consideration of nature development in the inner city
4. Historical continuity
5. Preservation of local diversity
6. Preservation of large interconnected open spaces
7. Interconnectivity of open spaces
8. Varying intensity of use
9. Preservation of diversity of typical elements in urban landscape
10. Prevention of all avoidable interferences with nature or landscape
11. Functional integration of built infrastructure into natural ecosystems
12. Creation of numerous air exchange corridors
13. Protection of all forms of life (Sukopp and Sukopp 1987, p 351–354)

In 2003, the organization DIE GRÜNE STADT (www.die-gruene-stadt.de) was established, which went on to become a foundation in 2009. Organizational forms such as this provide a platform for companies, individuals (including health experts), building managers and architects, homeowner associations, industrial companies, auditing firms, associations, Agenda-21 work-groups, municipalities and universities, all of which are interested in promoting more urban green. This results from the conviction that both urban green and urban nature have not been adequately represented/have been largely politically neglected and that there is a necessity for the exchange of knowledge and experiences, the creation and preservation of public and private green (spaces) in the city and a general improvement of nature awareness of citizens and decision-makers. The scope of activities ranges from indoor green and private gardens to the urban green of parks, botanical gardens and roadside green/vegetation. Nowadays, similar "Green City organisations" such as DIE GRÜNE STADT have been established in the Netherlands, Great Britain, France, Italy and Hungary and cooperate within the context of "Green City Europe." The concept of "green city" is currently expanding in Europe and is gaining broad public support (e.g. De Groene Stad (NL), De Goene Stad 2018, www.degroenestad.nl and The Green City (UK), The Green City 2018, www.thegreencity.co.uk) (DIE GRÜNE STADT 2018).

Furthermore, in many countries, broad alliances of clubs, organizations, foundations and companies also advocated urban green as a way of improving the quality of city life. National and international organizations of urban and spatial/development planners, NGOs and garden and landscape architects often lead such alliances.

Urban nature in the form of "urban green" can and should provide a greater contribution for sustainable urban development than it previously has. In this regard, eight focus areas can be identified:

1. Mitigating the effects of climate change
2. Improving public health
3. Securing social cohesion
4. Increasing real-estate value
5. Protecting soil, water and air
6. Preserving biodiversity
7. Promoting research in integration nature-based solutions into construction
8. Creating legal and financial incentives

Under the banner *Zukunft Stadt und Grün* (future city and green), politicians, administrators, economists, scientists and the general public are incentivized to cooperate and improve their efforts in regard to the development of urban nature (DIE GRÜNE STADT 2018, www.die-gruene-stadt.de).

Initially, the European National Strategies for biological diversity included urban nature only hesitantly. This was also the case with the German "National Strategy for biological diversity" (2007). In it, the field of action referred to as C 9 "settlements and traffic" covers all urban green and only mentions the use and fragmentation of land as priorities to be overcome.

At the same time, the demanded densification and increase in sealed surfaces often counteract this goal. In this regard, the primary goal for urban nature is the (necessary) restoration of the lost connection between the city residents and urban nature. Spaces for nature experiences and nature enjoyment should be reachable in short distances from home for all urban children (BMU 2007, p 79).

The "concrete visions" of the "National Strategies for Biological Diversity in Germany" describe either the current condition or are settling for very modest unaspiring concepts for urban greening. Among the general public the term "green" is typically used to summarize all notions of biodiversity in urban nature.

Our Vision for the Future Is Our cities have a high quality of life for city residents and also provide a habitat for many different species, including rare and endangered plants and animals. Diversified green spaces improve the air quality as well as the urban climate. It provides ample opportunities to relax, to play and experience nature for residents of all ages.

Our Goals To incorporate green infrastructure in residential areas including surrounding green spaces (e.g. courtyard green, small green spaces, roof and vertical green). Publicly accessible green spaces that offer a variety of functions and qualities are generally within walking distance and accessible (BMU 200, p 42).

On the initiative of the former German ministry of transport, construction and urban planning, interdisciplinary collaboration regarding the issue of "green in the city" was initiated in 2013. Its goal is to make the topic a political agenda and to initiate/encourage discussion processes. After the first German Federal Congress "Green in the City," which took place in 2015 in Berlin, the issue of urban green and its development was introduced as a future-oriented policy area and multifaceted political task. These issues were addressed in the *Grünbuch Stadtgrün, Grün in der Stadt. – Für eine lebenswerte Zukunft* (engl. Green Book Urban Green, Green in the City – for a Livable Future) (BmU 2015), which reflects the interdisciplinary approach of the federal government and considers not only the potentials but also possible areas of conflict. Hence, this document covers a broad spectrum of "urban green." On this basis, the German Ministry of Transport, Construction and Urban Development has initiated a board dialog on urban green.

In Germany and Europe, in general, the term "urban green" is currently being concretely defined and operationalized in both politics and urban planning. It encompasses all forms of urban nature, including both public and private lands (BBSR 2017, p. 8).

After the second German Federal Congress *Grün in der Stadt* (engl. "Green in the City") in 2017 in Essen, urban green was developed as a future-oriented policy area. A White Paper and recommendations of action soon followed/followed shortly after. The German government started the "German National Strategy for the Green City" in 2017 with a White Paper "Urban Green" of the German ministry ofEnvironment, Nature Conservation and Nuclear Safety and recommendations for ations (BMUB 2017).

Goals for action were defined for 10 different fields of action and made measurable via indicators, characteristic and guidance values. This is a great leap of progress regarding the pursuit of a green city concept. The White Paper *Stadtgrün* (engl. "urban green") is a proposal intended to support communities and other actors in their efforts to create, develop and maintain urban green as a self-evident aspect of integrative urban development.

The operational goals for the fields of action were (BBSR 2017) as follows:

- Climate and health
- Environment and nature
- Society and public space
- Organization and financing
- Urban space

3.2 Green Infrastructure – Local Basic Concept of Green City

The concept of "urban green Infrastructure" originates from urban planning. It was introduced to conceptualize the urban "green-system" as a coherent subject area of planning (Sandström 2002; Tzoulas et al. 2007).

Urban green infrastructure is a network comprised of semi-natural and designed elements of nature including green and open spaces and water bodies in cities. This includes nature in built-up and sealed areas. This network of different natural structures of varying size, location and ownership should be maintained and further developed as a communal responsibility of different political, economic and civilian actors. Regarding social, ecological and sustainable urban development, the goal for all elements of urban nature is as follows:

- To be accessible for all city residents
- Improve the health and well-being of city residents
- Enable both biological diversity and a range of possibilities to experience nature
- Contribute to the aesthetics of the city and simultaneously improve the quality of life within it
- Provide local ecosystem services for city residents

Urban green infrastructure significantly contributes to the quality of life and services for the public in cities. Sealed and built-up areas can become green urban infrastructure by means of unsealing surfaces, greening and planting trees.

At the EU level, "green infrastructure" (not necessarily in the urban context) is defined as a strategically planned European network on the supra-regional level. It is comprised of valuable natural, semi-natural and built-up areas, as well as further nature elements that provide important ecosystem services and contribute to the preservation of biodiversity (e. g. Dover 2015; Naumann et al. 2011; BfN 2017; BBSR 2017).

> *It (green infrastructure – the author) can be considered to comprise of all natural, semi-natural and artificial networks of multifunctional ecological systems within, around and between urban areas, at all spatial scale* (Tzoulas et al. 2007, p. 169).

The green city is based on green (and blue) infrastructure. This includes all waterways as part of urban nature; however, it is often further emphasized with the term "blue infrastructure." Although the main concepts of "green infrastructure" are not new, the term itself has become a popular expression only in the past decade.

By combining the terms "green" and "infrastructure," an attempt is made to endow urban nature with the same technological significance as built-up infrastructure and, thus, make the concept of urban green more assertive.

Infrastructure is a basic concept, without which the functioning of the urban complexity cannot exist. Consequently, this notion should also apply for "green" infrastructure – it is as much a necessity for the functionality of a city as is built infrastructure.

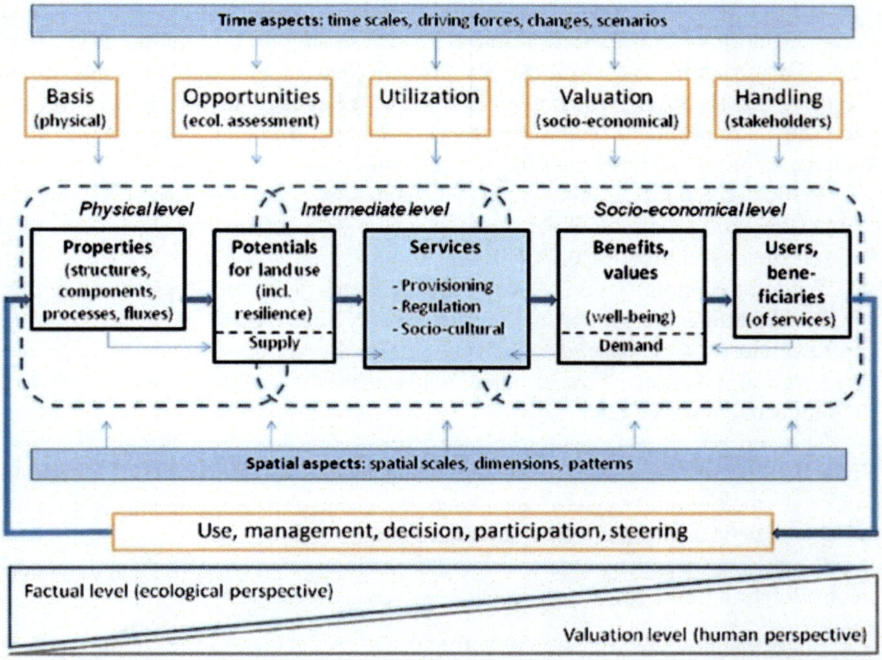

Fig. 3.2 Conceptual framework for the analysis of ecosystem services – the extended EPPS framework (Source: Bastian et al. 2013)

Green infrastructure or "green and blue infrastructure" defines a strategic planning network to promote nature on different scales.

The network of green infrastructures focusses on maintaining biodiversity as well as strengthening and improving the regenerative capacity of ecosystem and ecosystem services in terms of sustainable use of nature (Fig. 3.2).

Intensive land use and fragmentation of landscapes continue to threaten biodiversity on a global scale – especially in Europe. The concept of green infrastructure is intended to counteract this threat. It is an integrative approach intended to bring various actors together.

> *The term "green infrastructure" provides an opportunity to illustrate the public value of urban green, since the connotation with "infrastructure" is one of economic and social necessity* (Torsten Wilke, Municipality Leipzig, Department for Urban Green and Waters, cited by BfN 2017, p 5, translated).

On the supra-regional level (e.g. European scale), green infrastructure only relates to semi-natural and natural spaces. On the local urban scale, the term also includes all forms of urban infrastructure. Hence, there is a scale-based paradigm change. Instead of natural and semi-natural areas, the focus lies on "semi-natural and designed green spaces and elements" (BfN 2017, p 3). Hence, the new term of "urban green infrastructure."

Green infrastructure plays a particularly important role in urban areas. Here, the fragmentation of green spaces through surface sealing such as buildings and transport infrastructure and consequently also the loss of biodiversity is particularly high. However, through the implementation of the concept of "green infrastructure" in cities, a multitude of ecosystem services can be provided. For example, the air quality in cities can be greatly improved through parks and green spaces/areas.

Vertical greening can also greatly contribute by absorbing heat caused by sunlight, thus mitigating the adverse effects of the phenomenon referred to as "urban heat island" (Neßhöfer et al. 2012).

While in economics one typically differentiates between privately and publicly built infrastructure, this distinction does not typically occur in regard to "green infrastructure." Consequently, different actors, the public sector, private property owners and stakeholder, are included, which makes the development of networks a complicated and communal task.

The concept of urban green infrastructure represents strategic and integrative planning, protection, development and management of urban nature. This requires different district- and object-related spatial concepts for the entire city. This clearly exceeds traditional principles of urban planning.

Protection, management and development of urban green infrastructure occur in consideration of the following principles:

- Adapt usability and services of nature based on the needs of residents
- Develop strategic plans for said adaptation
- Connect nature
- Promote multi-usability and multi-functionality
- Allow uninterrupted/undisturbed development of nature and reduce maintenance if possible
- Include green infrastructure in areas with sealed surfaces and buildings
- Facilitate/advocate cooperation and alliances among different actors

The federal institute for Building, Urban Affairs and Spatial development (BBSR 2017) put central focus of its research on the following questions:

- Which systematic surveys for urban green and its quality exist?
- How do green areas and their accessibility develop, which trends are foreseeable?
- Which functions of urban green can be empirically derived?
- Which goal/objectives and standards for urban green in urban development exist today? Which standards for urban green can be systematically derived for communities?
- Which cities work towards goals involving urban green?
- Are there notions towards more urban green? If so, where and in which quality and what forms?
- How is the GALK (constant German Green Department Leaders Conference, defined standards for German urban green) list from 1973 to be evaluated from today's perspective? Is an update reasonable? Which recommendations for the standards of urban green can be developed?

- Which central, politically conveyable action goals can be derived?

The development of urban green infrastructure is primarily a task of the communities. They can line out for its suitable strategies and partnerships.

The understanding of urban nature as a system of internally integrated nature elements that interacts with its surroundings has been established. This system, when proactively planned and maintained as urban green infrastructure, has the potential to steer urban development and as such integrate economic growth, nature conservation and public health (Walmsley 2006; Schrijnen 2000; van der Ryn and Cowan 1996; Breuste et al. 2013). Hence, urban green infrastructure can be the key for developing the Green City. In the 1970s and 1980s, the basis for this concept was established by mapping biotopes in urban areas. This initially began in Germany but soon followed in other countries both within and outside of Europe (e.g. Japan and Brazil) providing a basis for comparisons among different cities.

The "Compact City within the Ecological Network" is the guiding principle of Dresden's urban planning towards becoming a Green City (Fig. 3.3).

The guiding principle encompasses compact urban settlements as settlement units, embedded in a network of ecologically functional spaces. The existing river system (400 community streams and the river Elbe) is the basis for the "ecological network" that will be gradually expanded together with green spaces. The following concrete functions are allocated to its specific nature areas:

- Clean air and a healthy urban climate

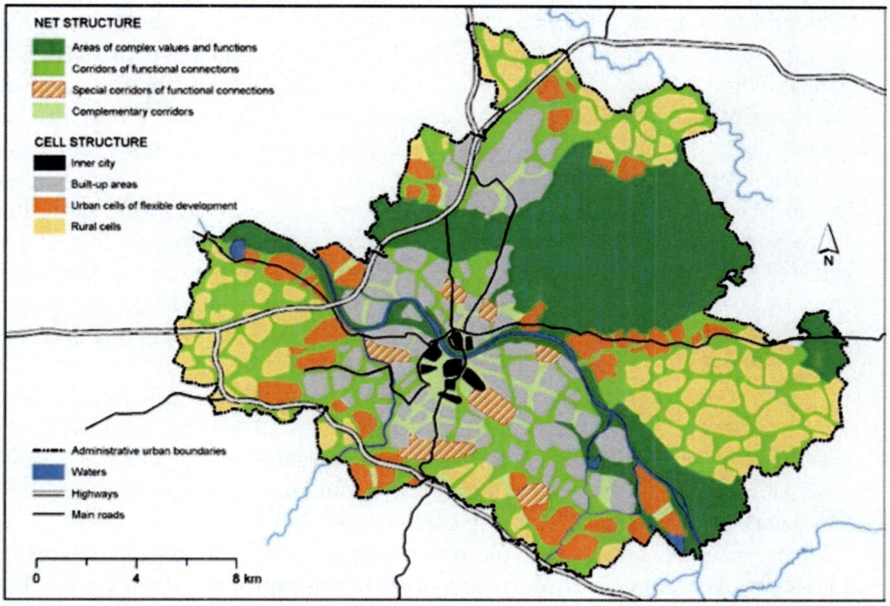

Fig. 3.3 The Green City Plan as landscape plan in Dresden, Germany (Source: REGKLAM 2015; Breuste et al. 2016, changed)

- Sufficient groundwater formation
- Flood prevention, water retention and development of water bodies
- Recreational areas for residents
- Habitats for plants and animals, as well as migration corridors and
- Beauty and uniqueness of cultural landscape

Measures for the development of water bodies (e.g. restoration and the vitalization of brownfields) strengthen the ecological network. Moderate development of built-up areas occurs in the defined spatial units, more intense development in the defined corridors leading into the city outskirts. The ecological network builds upon the basis of green corridors in the city. The implementation of this idea happens within a concept of development and measures. As such urban nature is perceived as infrastructure and open spaces as leading structures of urban development.

The necessary adaptation to climate change can be better managed. It requires more green spaces, as well as its interconnection within the urban landscape to mitigate the adverse effects of summer heat and heavy precipitation events (REGKLAM 2015; Breuste et al. 2016).

3.3 The European and Global Perspective

In 1994, European cities started an initiative for European cities and communities for a path towards a sustainable future. In 2004, a European process for sustainable development of cities by means of voluntary agreement was made concrete in Aalborg, Denmark, and included 2500 local and regional administrations from 39 countries and 80 European cities and communities. In Aalborg, approx. one thousand participants of the fourth conference for sustainable Cities and Towns, decided on the Aalborg Commitments (10 subject areas). Urban nature was not a central, yet integrated component. Objective number 3 "community natural assets: a commitment for accepting responsibility for the protection and preservation of natural community assets and ensuring their equal distribution" directly relates to urban nature. Task area 3 compels to "promote species diversity and to expand and maintain protected areas and green spaces." "Ecologically productive land" should also be preserved and "sustainable forestry" should be promoted, while "water, soil and air quality" should be improved (ESCTC 2004, 2013).

The Leipzig Charta 2007 recommends an "integrated urban development policy" as a process that simultaneously and equally considers all aspects and interests relevant for urban development. It focussed on innovation, competitiveness, public participation as well as balance, yet did not consider urban nature as a significant component for this objective (Leipzig Charta 2010).

Twenty European states are already members of the "European Landscape Contractors Association" (ELCA). In it, 74,000 companies and 330,000 members are represented. The concept of the Green City was developed in one of its first workshops in 2011.

For us the green city is the model of the future, creating urban structures with environments with life-quality. (ELCA 2011, p 4).

This model should help solve the anticipated problems for the following decades. This includes a new design of urban green, which combines water management, biodiversity and adaptation to climate change and health.

The four main principles for the Green City were identified as follows:

- Reconnection of cities with nature for the benefits of city residents.
- Urban nature must be an integral part of urban planning.
- Urban green is an interdisciplinary issue in regard to planning.
- Urban green should no longer be a neglected aspect of sustainability efforts.

In 2008, European scientists already urged municipalities, national governments and the European Commission to carry more responsibility regarding the implementation of Urban Green Strategies as a means to develop the Green City (Europäische Kommission/Generaldirektion Umwelt 2009).

Therefore, they call for the following.

3.3.1 City Authorities

- To develop, adopt, publicize and monitor a comprehensive Urban Green Space Strategy for their cities and to incorporate this strategy as a key element in the urban spatial and development planning
- To involve the different administrative departments of the municipality (influencing the urban green space development), scientists of various related disciplines, organizations, social groups and the general public in the preparation and implementation of this strategy with an especial emphasis in encouraging social inclusion
- To develop and expand on participation processes which build up trust, enhance cooperation between participants and encourage the public to take care of green spaces
- To develop local standards of quality, quantity and accessibility in order to permit different user groups to benefit, as appropriate, from urban green spaces
- To provide financial and personnel support for the departments responsible for the development and management of green spaces in order to permit them to plan, create and maintain them in accordance with local needs and conditions (including climatic, natural, social and economic)
- To cooperate with neighbouring cities and regions in the coordination of urban green strategies
- To create and develop related networks with other cities in Europe for the exchange of urban green space experience and the formulation of common policies and

- To invest in skills and focus on advanced training (lifelong learning) that relates to the relevance and management of urban green spaces

3.3.2 National Governments

- To support the implementation of more sustainable strategies for urban development at regional and local levels
- To raise awareness about the contribution of green space development to the ecological performance (biodiversity, climate change) of cities and support strategies to improve urban climate and ecological functions
- To introduce new financial resources for safeguarding and developing urban green spaces
- To incorporate appropriate finances for the development, adoption and monitoring of Urban Green Space Strategies
- To encourage the exchange of experiences between urban green departments and spatial planning departments nationally and across Europe and
- To take initiatives to secure a good data collection and data base to support Geographic Information Systems (GIS) applications

3.3.3 The European Commission

- To include the green space dimension in emerging EC Policy papers and initiatives to enhance city sustainability and competitiveness
- To encourage the development, adoption and dissemination of Green Space Strategies as a shared vision that meets the needs of urban dwellers and safeguards the natural and semi-natural spaces within the urban fabric
- To incorporate appropriate finances in the formulation, development and implementation of comprehensive Urban Green Space Strategies
- To finance special programmes (research and knowledge transfer) for the development of urban green spaces with an emphasis on enhancing biodiversity, pollution mitigation, climate change adaptation and improving public health
- To encourage the exchange of experiences between urban green departments and spatial planning departments across Europe and
- To promote the creation of new European organizations that support urban green space development (GreenKeys 2018, p 98–99)

The global perspective for sustainability in cities was already initiated in 1992 in Rio de Janeiro and made concrete for cities through the Habitat I–III conferences.

The Millennium Ecosystem Assessment Report (WRI 2005) also does not address cities as ecosystems and consequently also does not deal with urban nature nor provide a perspective for Green Cities. On a global scale, this issue was only approached via the conferences on habitat 3 and their preliminary meetings in 2016 in Quito.

In Habitat III, the New Urban Agenda for "Sustainable Cities and Human Settlements for all" was passed. It aims to create a unified vision and political responsibility for the promotion and implementation of sustainable urban development. A vision for Green Cities is not included in the principles and commitments of the New Urban Agenda. As far as the environment is concerned, resource efficiency regarding the resilience and ecological sustainability clearly prevails. The usual commitments for the protection of ecological resources and biodiversity remain vague and without concrete goals and instruments. Instead, risk management and resilience are clearly prioritized (United Nations 2016).

Within the work group "Urban Ecosystems and Resource Management," which had already compiled a strategy paper in New York in 2015, important keywords relating to the Green City can be identified; however, none of them were subsequently adopted into the New Urban Agenda. Some of these keywords included the following: Ecosystem services (ES), Ecosystem-based Adaptation (EbA), Green infrastructure (GI), also "preserving ecosystem-based management of cities, disaster risk reduction, health and recreation and even Citizens need to connect with nature, and benefit from this connection" (UN 2015, p 6).

Consequently, there is still no global vision for a Green City and its goals must give way to other issues such as poverty, inequality and an increase in necessary risk management (UN 2015).

Initially, the vision of the Green City always takes place on a local scale, which is where it belongs and where it can be implemented on an exemplary scale. The Green City is originally a European Concept, which has gained popularity, been further developed, implemented and supported on a global scale. The Green City provides direction in a crucial area, namely the relationship between nature and the city. It therefore remains a concise and implementable concept that can be gradually executed. It is an essential part of the more complex and multifaceted vision of a sustainable city.

References

Bastian O, Syrbe R-U, Rosenberg M, Rahe D, Grunewald K (2013) The five pillar EPPS framework for quantifying, mapping and managing ecosystem services. Ecosyst Serv 4:15–24

Breuste J, Endlicher W (2017) Stadtökologie. In: Gebhardt H, Glaser R, Radtke U, Reuber P (eds) Geographie – Physische Geographie und Humangeographie, 3rd edn. Elsevier, München, pp 628–638

Breuste J, Pauleit S, Pain J (eds) (2013) Stadtlandschaft – vielfältige Natur und ungleiche Entwicklung. Schriftenreihe des Kompetenznetzwerkes Stadtökologie (= Conturec 5). Darmstadt

Breuste J, Pauleit S, Haase D, Sauerwein M (eds) (2016) Stadtökosysteme. Springer, Berlin und Heidelberg

Bundesamt für Naturschutz (BfN) (eds.) (2017) Urbane grüne Infrastruktur. Grundlage für attraktive und zukunftsfähige Städte. Hinweise für die Kommunale Praxis, Bonn

Bundesinstitut für Bau-, Stadt- und Raumforschung (BBSR) (eds) (2017) Handlungsziele für Stadtgrün und deren empirische Evidenz. Indikatoren, Kenn- und Orientierungswerte, Bonn
Bundesministerium für Umwelt, Naturschutz, Bau und Reaktorsicherheit (BmU) (2007) Nationale Strategie zur biologischen Vielfalt. Vom Bundeskabinett am 7. November 2007 beschlossen. Reihe Umweltpolitik. Berlin
Bundesministerium für Umwelt, Naturschutz, Bau und Reaktorsicherheit (BmU) (2015) Grün in der Stadt – Für eine lebenswerte Zukunft. Grünbuch Stadtgrün, Berlin
Bundesministerium für Umwelt, Naturschutz, Bau und Reaktorsicherheit (BMUB) (eds) (2017) Grün in der Stadt – Für eine lebenswerte Zukunft. Berlin
De Groene Stad (NL) (2018) De Groene Stad. www.degroenestad.nl. Accessed 5 Jan 2018
Die Grüne Stadt (2018) Die Grüne Stadt. www.die-gruene-stadt.de. Accessed 5 Jan 2018
Dover JW (2015) Green infrastructure. Incorporating plants and enhancing biodiversity in buildings and urban environments. Earthscan/Routledge, London/New York
Europäische Kommission/Generaldirektion Umwelt (eds) (2009) Zielrichtung: eine grüne Infrastruktur in Europa. In: Natura 2000: Newsletter "Natur und Biodiversität" der Europäischen Kommission. 27:3–7
European Landscape Contractors Association (ELCA) (eds) (2011) ELCA research workshop green city Europe for a better life in European cities. www.green-city.de. Accessed 5 Jan 2018
European Sustainable Cities & Towns Campaign (ESCTC) (2004) Die Aalborg commitments. www.ccre.org/docs/Aalborg03_05_deutsch.pdf. Accessed 7 January 2018
European Sustainable Cities & Towns Campaign (ESCTC) (2013) European Sustainable Cities. www.sustainablecities.eu. Accessed 12 Jan 2014
GreenKeys (2018) Recommendations for new urban green policies and an agenda for future action. www.greenkeys-project.net; greenkeys@ioer.de. Accessed 7 Jan 2018
Leipzig Charta (2010) Leipzig-Charta zur nachhaltigen europäischen Stadt 2007. Informationen zur Raumentwicklung 4:315–319
Naumann S, McKenna D, Kaphengst T, Pieterse M, Rayment M (2011) Design, implementation and cost elements of green infrastructure projects. Final report to the European Commission, DG Environment, contract no. 070307/2010/577182/ETU/F.1. Ecologic institute and GHK Consulting, Brüssel
Neßhöfer C, Kugel C, Schniewind I (2012) Ökosystemleistungen im Europäischen Kontext: EU Biodiversitätsstrategie 2020 und "Grüne Infrastruktur". In: Hansjürgens B, Neßhöver C, Schniewind I (Hrsg) Der Nutzen von Ökonomie und Ökosystemleistungen für die Naturschutzpraxis. Workshop I: Einführung und Grundlagen. In: BfN-Skripte 318, p 22–27
REGKLAM Regionales Klimaanpassungsprogramm für die Modellregion Dresden (2015) Grüne und kompakte Städte. www.regklam.de/...programm/.../gruene-und-kompakte-staedte/?tx.... Accessed 15 Apr 2015
Sandström UF (2002) Green infrastructure planning in urban Sweden. Plann Pract Res 17(4):373–385
Schrijnen PM (2000) Infrastructure networks and red-green patterns in city regions. Landsc Urban Plann 48:191–204
Sukopp H, Sukopp U (1987) Leitlinien für den Naturschutz in Städten Zentraleuropas. In: Miyawaki A, Bogenrieder A, Okuda S, White J (eds) Vegetation ecology and creation of new environments. Tokai University Press, Tokyo, pp 347–355
The Green City (UK) (2018) www.thegreencity.co.uk. Accessed 5 Jan 2018
Tzoulas K, Korpela K, Venn S, Yli-Pelikonen V, Kazmierczak, Niemela J, James P (2007) Promoting ecosystem and human health in urban areas using green infrastructure: a literature review. Landsc Urban Plann 81:167–178
United Nations (UN) (2015) Habitat III issued papers. 16 urban ecosystems and resource management. New York. www.habitat3.org. Accessed 18 June 2018
United Nations (UN) (2016) Habitat III. Neue Urbane agenda. www.habitat3.org. Accessed 18 June 2018
van der Ryn S, Cowan S (1996) Ecological Design. Island Press, Washington, DC

Walmsley A (2006) Greenways: multiplying and diversifying in the 21st century. Landsc Urban Plann 76:252–290

World Resources Institute (WRI) (2005) Millennium ecosystem assessment. Washington. http://www.millenniumassessment.org/documents/document.356.aspx.pdf. Accessed 18 June 2018

Chapter 4
The Inertial Forces of Ecological Planning: How Planning Resists Conceptual Change

Kimmo Lapintie

Abstract Spatial planning – in both its ideas and practices – cannot directly adopt new concepts, but instead they must be embedded in the existing framework of professional concepts. This is challenging for planning discourses when determining the ways in which the urban green is conceptualized. This chapter argues that the basic dichotomy of recreational and preservation values has been able to resist the introduction of new concepts, such as green infrastructure and ecosystem services. This resistance can be understood through an 'archaeological' analysis of planning discourses, using Foucault's analysis of the discourse formations of serious speakers. In addition to the usual dichotomies of urban growth versus green or the built environment versus nature, there is a deeper dualism between rationalized nature and the 'bestiality' of uncontrolled and uncommunicated nature, which is still needed for the legitimacy of planning. Current conceptual frameworks in the planning of the cities in Helsinki, Milan, and Montreal are analysed from this perspective.

Keywords Ecosystem services · Green infrastructure · Planning discourse · Foucault · Archaeology of knowledge · Recreation · Preservation · Milan · Montreal · Helsinki

4.1 Introduction

During recent decades, spatial planning has been confronted with a host of new concepts, inspired by the growing environmental awareness and, particularly, the concept of sustainable development from the late 1980s (Di Marino et al. 2018; Haaland and van den Bosch 2015; Barbosa et al. 2007; Brundtland 1987). The most recent of these, such as green infrastructure, ecosystem services, and nature-based

K. Lapintie (✉)
Department of Architecture, Helsinki School of Arts, Design and Architecture, Helsinki, Finland
e-mail: kimmo.lapintie@aalto.fi

A. Arcidiacono, S. Ronchi (eds.), *Ecosystem Services and Green Infrastructure*, Cities and Nature, https://doi.org/10.1007/978-3-030-54345-7_4

solutions, are challenging the way in which urban green areas have traditionally been conceptualized. This conceptualization has been in terms of a passive background, on top of which the active development of cities and regions are developed, or as selected green patches of natural or recreational areas that should be protected or constructed. It is eventually expected that a more holistic and dynamic conception of urban ecology will, in the end, produce new planning practices that will help to more efficiently address environmental problems, as well as produce a more pleasant living environment for citizens.

However, in this chapter, I argue that spatial planning – in both its ideas and practices – cannot directly adopt new concepts, but they have to be embedded in the existing framework of professional concepts. Concepts, contrary to what one might first think, are not simply representations of reality: there is no 'city', 'urban fabric', 'green structure', or even 'ecology', to which we could refer to. Rather, they are conceptual tools, something that we need in order to make sense of reality and our own activities. As such they are artefacts; they have been created and adopted, but they can also be changed. New concepts will arise, but there needs to be a coherence – in terms of logic, but also in terms of value – between the existing concepts. These have their own history and affiliation with the professional identity, as well as social and political legitimacy.

Although urban and regional planning is often perceived as a practical discipline, it is impossible to circumvent this conceptual or even philosophical problem. However, in what sense is it a problem? Since 1993, we have studied the concepts used in 'ecological planning' – planning that consciously addresses issues such as sustainability, green-blue structure or ecosystem services – with varying groups of researchers. We noticed in the very beginning – when debates on sustainability and ecology in planning were still in their infancy in the Finnish context that we studied – that two concepts clearly dominated the planning discourse: recreation and preservation. In a sense, they were in opposition to one another, and both of them were naturally in conflict with the shift of urban land use to more 'economic' uses, such as for housing or industry. Recreation referred mostly to human use of green areas in the city (which are currently called 'cultural ecosystem services'), while preservation (in the meaning of protecting existing values of nature) was about minimizing human activities that might endanger any rare species or landscapes, including the impacts of extensive recreational use. There was no way to 'objectively' judge the amount of urban space that should be left for recreation or the kinds of natural values which were worth protecting; this essentially remained a political question.

Much has happened since then: new concepts, such as ecosystem services, green infrastructure, or nature-based solutions, have gained a foothold in planning discourses, and the keyword of sustainability, which was still fresh in the early 1990s, has become a must-have in every planning text around the world. A growing body of literature on ecological planning is available for planners wishing to assess the sustainability or environmental impacts of their plans. Yet, what has actually changed? In our recent studies (Di Marino and Lapintie 2017; Di Marino et al. 2018, 2019; Lapintie and Di Marino 2019), we have discovered that there is a conceptual variety and ambiguity in different cities, but two concepts of recreation and preservation continue to dominate the scene in several cases.

Let us present a concrete example from these studies. In the new comprehensive plan for the city of Helsinki, Finland (approved by the City Council in 2016), the concept of a 'network of green areas' was used. They were said to include 'many scenic and historic entities, large recreational areas, such as outdoor sports parks and neighbourhood parks, as well as protected areas, such as Natura areas'. Thus, the categorization was still based on the traditional dichotomy, recreation and preservation of nature. These two main categories were also represented in two separate thematic maps on 'recreational network' and 'urban nature' (Di Marino and Lapintie 2017).

This is a phenomenon which I shall call 'the inertial forces of ecological planning'. It means the stubbornness of the planning discourse in resisting any conceptual changes which might be on offer, if they do not quite fit the existing and traditional framework to which one is accustomed. Understanding these forces appears to be necessary if we want to influence the way that planners think and act.

We must ourselves rid of the assumption that new concepts and new knowledge would somehow automatically become part of planning knowledge and practice. Instead, we have to analyse the way that discourses are formed in the professional and political contexts in which planners are operating, as well as the points at which their thoughts and acts meet those of other stakeholders of urban development.

4.2 An Archaeology of Silence

Understanding planning discourses becomes challenging as soon as we surrender the 'natural' attitude of referring to the way things 'really' are or even how they seem to be, thereby 'decentring' both the object and the subject. Without necessarily denying the obvious connection between discourses and their non-discursive contexts, it seems plausible to practise a sort of 'double bracketing': trying to perceive the discourses themselves as professional practices. In this way, we may understand the 'rarity' that professional discourses exemplify: what can be said and what cannot be said by serious speakers (Foucault 1985b). This will also let us address the problem mentioned above: why do planning discourses not change along with the change in other disciplines (such as environmental sciences), despite the state of the world also changing and presenting new challenges, such as climate change.

This 'archaeological method' was developed by Michel Foucault in his earlier writings, particularly *Les mots et les choses (une archéologie des sciences humaines)* from 1966 and *L'Archéologie du savoir* from 1969. It was a daring attempt to develop a theory and methodology of discourse formations as independent, rule-governed structures. It is based on an analysis of historical discursive formations, in which statements are understood as events, and the relationships of these statements are studied inside the discourses. This does not mean that they would be unrelated to non-discursive elements, but the way they are formed is not based on a direct reference to transcendental objects (Foucault 1985b, p. 49) or the experience or meaning-giving of a subject (ibid. p. 54).

Thus, Foucault was using double-bracketing: without denying the existence (or even the relevance) of non-discursive events, he sought the rules that determine the formation of statements within a group of serious speakers. Unlike language (langue) which allows an infinite set of possible sentences, the discursive formations are characterized by rarity (ibid. p. 118): not everyone has the authority to speak seriously (about medicine, about economics, about science, about planning and so on), and there is a rarity in what they can say.

> *This authority also involves the rules and processes of appropriation of discourse: for in our societies (and no doubt in many others) the property of discourse – in the sense of right to speak, ability to understand, licit and immediate access to the corpus of already formulated statements, and the capacity to invest this discourse in decisions, institutions, or practices – is in fact confined (sometimes with the addition of legal sanctions) to a particular group of individuals; in the bourgeois societies that we have known since the sixteenth century, economic discourse has never been a common discourse (no more than medical or literary discourse, though in a different way).* (ibid. p. 68)

However, in their famous critique of the archaeology of knowledge, Hubert Dreyfus and Paul Rabinow argued that this double-bracketing is a problem for Foucault, since there will be no consistent way of grounding the rules of formation that are so essential to archaeology (Dreyfus and Rabinow 1983).

> *The discursive practices analysed by the archaeologist are motivated by the speakers' conviction that they are uttering serious truths about man and society, or that they are helping to make explicit the implicit thoughts of those who were in possession of such truths. The analysis, however, substitutes for this 'naive' conviction as its condition of occurrence a set of meaningless strict rules.* (Dreyfus and Rabinow 1983, p. 93–94)

This critique has been taken for granted by many scholars, which is a pity, according to Tuomo Tiisala (2015), because it has prevented research in the humanities and social sciences from developing archaeology into a fruitful methodology. Tiisala argues that Dreyfus and Rabinow have not understood the pragmatic turn in Foucault's thinking, which enables the assumption of strict rules of discourse formation that are not accessible to the speaker's consciousness, but which they learn through practice. He refers to the unpublished manuscript of the Archaeology of Knowledge, in which Foucault still defined the rules of discourse formation as statements, which would indeed have undermined his idea of unconscious rules that are followed (but not known) by the speakers (Tiisala 2015, p. 659). Nevertheless, in the published version of the Archaeology of Knowledge, Foucault clearly states that the rules are part of the discursive practices:

> *By system of formation, then, I mean a complex group of relations that function as a rule: it lays down what must be related, in a particular discursive practice, for such and such an enunciation to be made, for such and such a concept to be used, for such and such a strategy to be organized. To define a system of formation in its specific individuality is therefore to characterize a discourse or a group of statements by the regularity of a practice.* (Foucault Michel Foucault 1985b, p. 74)

Archaeology of knowledge is also suitable for analysing the planning discourse, despite the early Foucault not discussing it. This raises the question, can one analyse – by using the archaeological method – that which is not said? This seems to be crucial in analysing contemporary discourses. In *Madness and Civilization*, Foucault clearly had this ambition: "The language of psychiatry, which is a monologue of reason *about* madness, has been established only on the basis of such a silence. I have not tried to write the history of that language, but rather the archaeology of that silence". (Foucault 1973, p. xi, emphasis in the original). However, in the *Archaeology of Knowledge*, he seems to be saying almost the exact opposite:

> *We are studying statements at the limit that separates them from what is not said, in the occurrence that allows them to emerge to the exclusion of all others. Our task is not to give voice to the silence that surrounds them, nor to discover all that, in them and beside them, had remained silent or had been reduced to silence. Nor is it to study the obstacles that have prevented a particular discovery, held back a particular formulation, repressed a particular form of enunciation, a particular unconscious meaning, or a particular rationality in the course of development; but to define a limited system of presences. The discursive formation is not therefore a developing totality, with its own dynamism of inertia, carrying with it, in an unformulated discourse, what it does not say, what it has not yet said, or what contradicts it at that moment; it is not a rich, difficult germination, it is a distribution of gaps, voids, absences, limits, divisions* (Foucault 1985b p 119).

How then can you define the limits of discursive formation without going to the 'other side' and describing it? Does this in the end mean providing silence with a voice? This difficulty reminds us of the famous observation by Wittgenstein that seeing the limits of your world and language (which are the same limits) is not possible, since you would in a way need to 'measure' them from the outside (Wittgenstein 1961, 5.61). Nevertheless, the limits to which Foucault refers are less all-encompassing; they are the limits of the group of experts that have the authority (at a certain historical time and place) to speak seriously about topics, such as madness, economy, science, and planning.

Foucault himself had the benefit of following the longer span of history to find the gaps and absences. In the *Madness and Civilization*, he pointed out the lost dialogue between the men of reason and the men of unreason (*folie*) that still existed during the time of Cervantes' Don Quixote and Shakespeare's Macbeth; this 'silence' we can still read from their texts. On the other hand, in *The Order of Things*, he used Aldrovandi's *History of Serpents and Dragons* as an example of the time (in the sixteenth century), when the observed facts and fiction had not yet been separated, and compared it to Jonston's *Natural History of Quadrupeds*, in which the fiction was simply left out (Foucault 1985a p. 128). This did not mean that Jonston would have known *more* (actually he knew *less*), but the two books represented the transition from similarity (serpents and dragons are indeed similar) to representation. A biologist of today – or any other author of non-fiction – could never write a book on serpents *and* dragons. In this case, too, the absence could be seen in the historical change in the scientific discourses.

However, if we want to use the archaeological method to study the contemporary formation of statements, we need to construct the 'dragons' to locate the limits of the professional discourses. We should not simply say that planners do not talk much about the 'other' dimensions of urban nature (e.g. besides recreation and preservation), or that they *should* do that; rather, we should try to understand the reasons behind contemporary planning discourse being as it is; the reasons for the limits of discourse being drawn in this particular way, thus creating this particular rarity. Hence, the analysis must concentrate on the rules within the discourse, revealing its hidden logic. However, the presence of something is the absence of something else. Even if the discourses would not hide or repress alternative discourses, planning research (unlike planning practice) can also try to measure the limits of the discourses of practice by viewing them from the outside. In this way, we may indeed discuss their 'dynamisms of inertia'.

4.3 Beauty and the Beast

Instead of a detailed analysis of the various planning discoursers used in different cities, I will try to make sense of the observation that a certain conceptual framework (such as the dominance of recreation and preservation) can continue to maintain its dominant position in planning discourse in spite of the challenge of new concepts, such as sustainability, ecosystem services, green infrastructure or nature-based solutions. Often these concepts are mentioned, but they have not been able to unbalance the traditional framework. For instance, in the master plan of Helsinki that we analysed (Di Marino and Lapintie 2017), the sentence 'Ecosystem services should be developed' can be found as a general requirement of the plan, but none of its implications can be found in the text or in the visual representations used.

Similarly, in two other case studies that we studied (Di Marino et al. 2019), the traditional representations, such as maps, seemed to be ill-suited to addressing the dynamic features of ecosystem services provided by the green infrastructure. They were more suitable for dividing land for different purposes, such as recreation or housing, and also for safeguarding some of the most valuable natural areas from exploitation. How is this to be understood? Clearly new forms of representation (such as GIS) can help in this respect, but there is clearly more to it than that. We should not assume that a new medium would automatically change the message in this case.

Returning to the original dichotomy between recreation and preservation, in some sense, it represents a functional/non-functional division. In the planning of urban land use, the functional divisions naturally dominate the conceptual framework: land is allocated to the human purposes of housing, industry, public and private services, transportation – and recreation. In a functional city, one should be able to recreate oneself, in order to remain as part of the productive workforce. As the urban green is known to have such restorative effects (Tyrväinen et al. 2014; Van den Berg et al. 2014), a sufficient measure of accessible green areas should be provided for urban citizens. However, since recreational use is also a potential disturbance to the ecological features of these areas, it is important to safeguard the most important natural values by managing and partly also preventing their functional use. Furthermore, typical urban uses of green areas, such as jogging, walking, and enjoying the scenery, require active construction of suitable features, such as pedestrian paths, benches, and lawns, which also correspond to the aesthetic preferences of the urban population. To a great extent, the recreational urban green is also an artefact, a 'bird in a cage'; in a way, it represents the original meaning of ecology (oikos-logos), as rational housekeeping.

However, how is this related to preservation? Perhaps, by stating that it is the opposite of recreation, being something that is safeguarded, left as it is, distinguished from functional purposes including extensive recreational use. It may be lightly managed, and in practice the borders between recreational and preserved areas may be blurred, but conceptually there is a difference. Since there is no human function allocated to it, its status as a protected green area has to be determined in another way. The rational (sic) behind it may be our attempt to let 'nature have its way', at least in some selected patches of our environment. This conceptual dichotomy can also be reflected in representations that define the limits of protected areas, excluding construction, as well as sometimes even excluding humans from visiting the area. However, are we talking about nature as being distinct from us and our activities?

This is an issue which I discussed in an earlier article (Lapintie 2005). Urban planning operates in a series of dichotomies, such as urban growth and green, red and green, development and preservation, or man and nature. However, there lies behind these a deeper dualism. As already mentioned, the apparent opposites of rational activities – urban green, preserved areas, recreational areas – are also objects of rationalization: they are constructed, managed, contained, valorized. Whatever freedom is given to them, it is indeed *given*, as part of our strategy to develop an urban green structure providing us with different ecosystem services, at least biodiversity which is necessary for human beings in the long run. Rationality – in spite of the legacy of the critique of rational planning – seems to be the winning strategy.

For instance, the city of Newcastle unsurprisingly calls the green corridor along the Tyne river a 'Strategic Green Infrastructure Network', which is composed of six main corridors linking important green spaces. Indeed, concepts, such as green infrastructure or ecosystem services, have been introduced to highlight the importance of the urban green to citizens, comparable to the 'hard' infrastructure of roads, ducts and wires, as well as the public and private services of the welfare state.

Yet, in order to make sense, planning must have its counterpart, something that legitimizes it. If it is not nature, if nature is already strategically caged and managed, what is it? Where is the 'beast'? Bestiality is indeed a conceptual element that threatens the human condition, and what is more: we cannot communicate with it.

Thus, Habermas' communicative rationality is equally armless as instrumental or strategic rationality in front of the beast. It can only be contained, silenced or destroyed. It cannot be given a voice, not through an archaeological or any other method. Jacques Derrida saw this paradox quite clearly:

Is not an archaeology, even of silence, a logic, that is, an organized language, a project, a sentence, a syntax, a work? (Derrida 1985, p. 35). And further: *Since the revolution against reason, from the moment it is articulated, can operate only within reason, it always has the limited scope of what is called, precisely in the language of the department of internal affairs, a disturbance.* (ibid p. 36)

4.4 Planning for the State Apparatus

Since planning is undeniably part of the state apparatus, it naturally aims at totalization: creating an epistemology of all that needs to be governed, as well as policies and technical infrastructure that will produce the desired effects. However, in their *Thousand Plateaus*, Deleuze and Guattari convincingly argued that there is always an exteriority that is not part of the state apparatus and not captured by the 'Royal Science' ('war machine' or 'nomadism' as they call it)(Deleuze and Guattari 2019). On the other hand, as argued by Foucault, the planning discourse does not address its exteriority.

Let us consider the cities of Helsinki, Milan and Montreal, which we have earlier discussed (Di Marino and Lapintie 2017). In the comprehensive City Plan of Helsinki, the green structure of Helsinki is represented as continuous green fingers (*vihersormet*), green connections (*viheryhteydet*) and green lines (*viherlinjat*), as

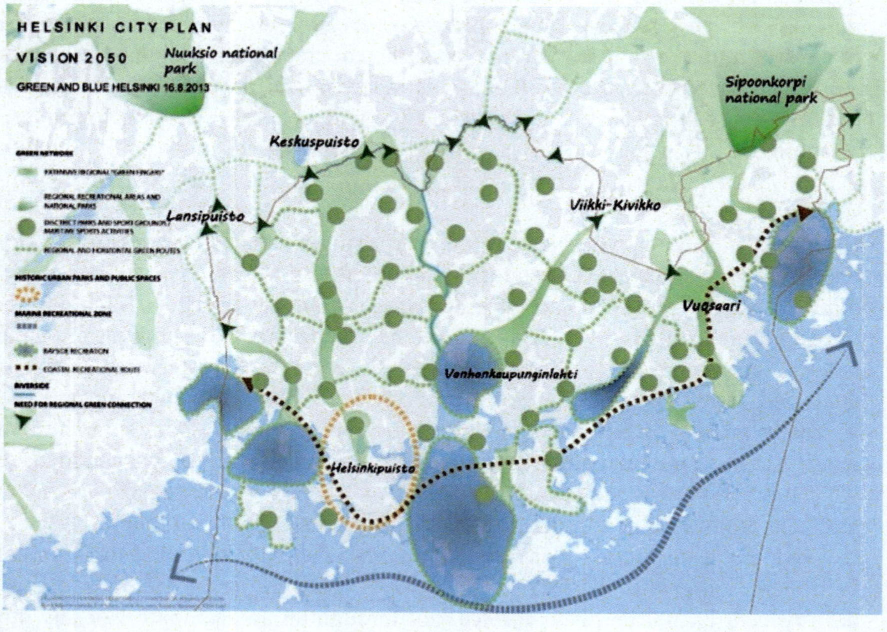

Fig. 4.1 Green and blue Helsinki (Source: Helsinki City Plan, Vision 2050, City Planning Department of Helsinki (City of Helsinki 2013))

well as larger green areas and neighbourhood parks (Fig. 4.1). The continuity is mainly functional, since the fingers and lines are broken in several locations by roads and rail tracks, only allowing a recreational connection through pedestrian underpasses or bridges.

Of course, what is noteworthy is that a representation could be drawn in which the built-up areas are white and the 'hard infrastructure' left out. Although the borders are blurred, there is an implicit dualism between red (or grey) and green. This is of course inconsistent with the concept of a green infrastructure providing ecosystem services which are not confined to the un-built areas. Urban floods do not respect administrative borders, stormwater gathers in built-up areas as well, and there can be more or less biodiversity in many types of green areas from golf courses to urban forests, none of which are categorized in the plan. The gardens of residential areas are also part of the green infrastructure. The ecological features of the city are more or less superimposed rather than contained within the green areas.

It is also noteworthy that the final plan – as a map, as a drawing – does not have any dynamic features. A loss of biodiversity, frequency and severity of flooding and the role of the urban green to mitigate them, erosion as a result of construction and increased use, or adaptability to heat waves and poor air quality are not addressed – and cannot be addressed with a blueprint or snapshot of the city in 2050.

The model used in the territorial plan of the Milan metropolitan area is in many respects similar (Fig. 4.2). The aim is to draw an ecological network (*rete ecologica*) that is composed of larger green areas, primary and secondary core areas (*gangli*) and primary and secondary green or blue corridors (*corridoi terresti e corridoi fluviali*). The primary and secondary core areas around the city of Milan are larger than in Helsinki, but the city itself is more compact, thus remaining mainly outside the dedicated ecological network. The borders between the green and the 'white' are clearer, and there is also a clearer division between the city and the countryside, representing the different cultures of urbanism in the two countries.

Whereas both Helsinki and Milan highlight the continuity of the ecological network, the city of Montreal concentrates on individual patches of green; understandably, since most of the land has already been used for residential and industrial purposes (Fig. 4.3). It would now be difficult or impossible to connect the parks and other green areas into a network anymore. Therefore, protection and conservation inside the perimeter of the remaining green areas are highlighted, with even 'ecoterritories' being specified. 'The Regional Plan identifies the Urban Agglomeration's main areas of interest: built and archaeological heritage, sectors of ecological importance and iconic landscapes. It proposes various measures to ensure their protection and promotion'.

In spite of the national vocabulary and connectedness, one could conclude that planning is still very much territorial dedicating green areas mainly to recreation and protection, based on their initial ecological values, as well as those that can be promoted in particular through connectivity. The ecological functioning of the green infrastructure is not superimposed on the other functions, which is in line with the functionalist (and real estate) tradition in land division.

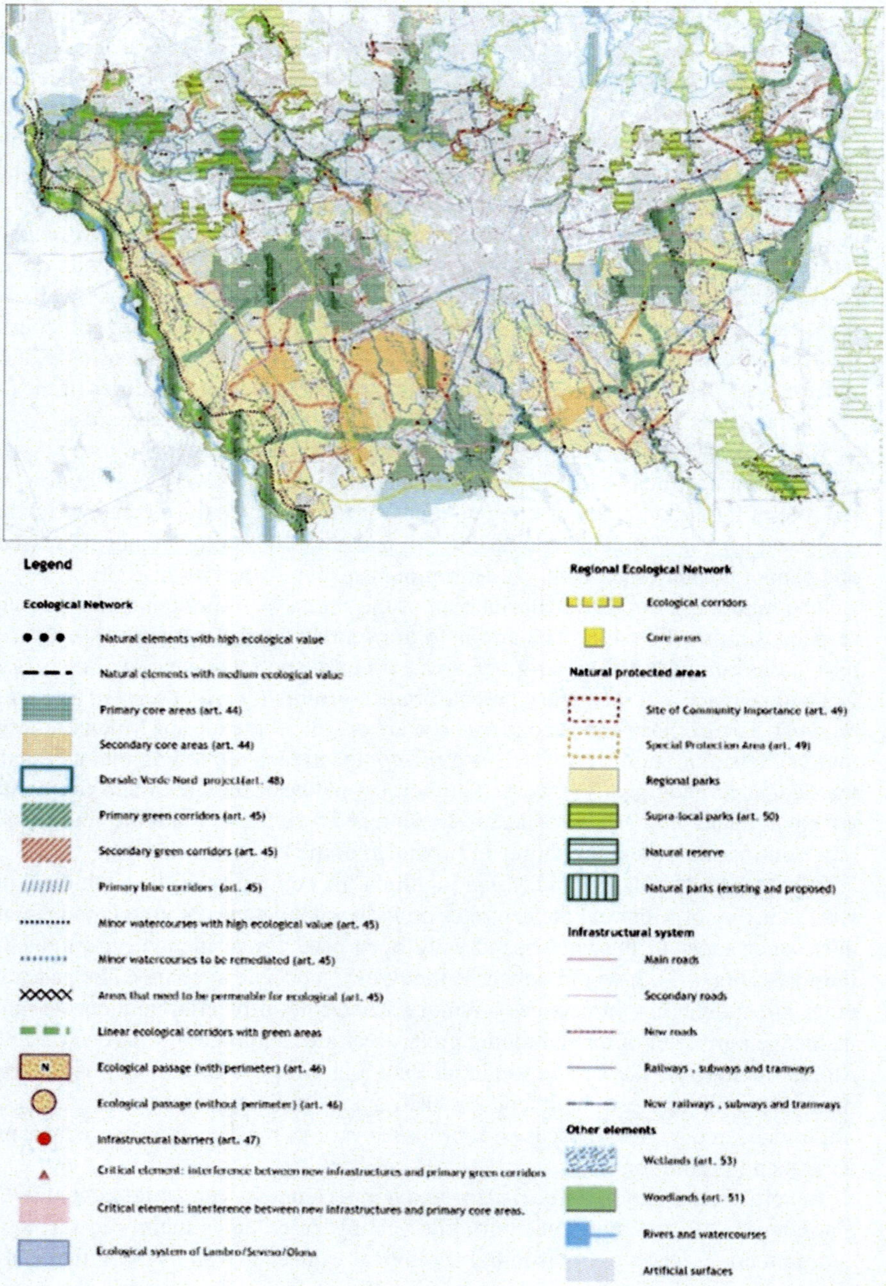

Fig. 4.2 Ecological network of the territorial plan of the Province of Milan (2013) (now Milan Metropolitan area) (Source: Piano territoriale di coordinamento provinciale della Provincia di Milano (Province of Milan 2013))

Fig. 4.3 Spatial organization concept of the Montréal Urban agglomeration land use and development plan (2014) (Source image: Montréal Urban agglomeration land use and development plan – draft September 2014 (City of Montréal 2013))

4.5 Beyond the Limits

Nonetheless, if ecosystem services and green infrastructure, as well as recreation and preservation, are equally concepts *within* rationality, how can we explain the inertial forces of planning that we started with? Planning and its epistemology are naturally rational processes, part of the state apparatus; hence it would seem easy to extend the rationalization from contained areas of recreation and preservation to an analysis of the whole green infrastructure, producing the various ecosystem services with respect to our human condition: provisioning (food, water and materials to *us*), regulating (*our* climate and keeping *our* feet dry), supporting (nutrients and oxygen to *us*) and cultural (keeping *us* happy and productive). It is natural that this conceptual framework is based on human interests, and it is exactly this emphasis that is expected to provide force to it. Rationality aims at a totality, a system of governance which takes care of everything.

Regardless of this, on closer inspection, we can see a crack, or a gap, that will break this sealed system. Although it is possible to govern life inside a specified area (analysing and supporting its values and ecosystem services, changing only functionalism to multifunctionalism), it is clear that this system is never closed.

Perhaps containment is the necessary condition of all 'departments of internal affairs'. However, as soon as global connectivity and international dependence are realized, planning will indeed meet its beast: the unmanageable path of the climate change, and the weakness of the global political system to respond to it. These are no longer just internal affairs.

Therefore, the concepts of green infrastructure and ecosystem services can open the door to the necessary counterpart of rational planning, the *folie*: something that we cannot control or understand, and something that defies any attempt to engage in a rational discussion with it. This idea has a long history in Western thought, from Plato's idea of the *wise* people who should govern the majority that was led by its *desire*, and not only its desire but also its will-to-power. Nevertheless, this dualism is still needed to provide legitimacy for planning as a control mechanism, and it is this challenge that planning must face, even at the price of losing its traditional identity. Climate change and a loss of biodiversity have replaced the diseases and immorality from which modern planning originally derived its legitimacy. They were perhaps solved, but a new beast has raised its head.

References

Barbosa O, Tratalos JA, Armsworth PR, Davies RG, Fuller RA, Johnson P, Gaston KJ (2007) Who benefits from access to green space? A case study from Sheffield, UK. Landsc Urban Plan 83:187–195

Brundtland GH (1987) Report of the World Commission on environment and development: our common future. United Nations, Oslo

City of Helsinki (2013) Helsinki City Plan, Vision 2050, Urban plan – the new Helsinki city plan. http://www.hel.fi/hel2/ksv/julkaisut/yos_2013-23_en.pdf

City of Montréal (2013) Montréal Urban agglomeration land use and development plan – draft September 2014. http://ville.montreal.qc.ca/pls/portal/docs/PAGE/COMMISSIONS_PERM_V2_FR/MEDIA/DOCUMENTS/DOCCONSULT_SUMMARY_AN_20140922.PDF

Deleuze G, Guattari F (2019) A thousand plateaus. Bloomsbury, London

Derrida J (1985) Writing and difference. Schocken, New York

Di Marino M, Lapintie K (2017) Exploring the concept of green infrastructure in urban landscape. Experiences from Italy. Canada Finland Landsc Res 43(1):139–149. https://doi.org/10.1080/01426397.2017.1300640

Di Marino M, Niemelä J, Lapintie K (2018) Urban nature for land use planning: the City of Helsinki. Urbanistica 159:94–102

Di Marino M, Tiitu M, Lapintie K, Viinikka A, Kopperoinen L (2019) Integrating green infrastructure and ecosystem services in land use planning. Results from two Finnish case studies. Land Use Policy 82:643–656

Dreyfus HL, Rabinow P (1983) Michel Foucault: beyond structuralism and hermeneutics. The University of Chicago Press, Chicago

Foucault M (1973) Madness and civilization. A history of insanity in the age of reason. Vintage Books, New York

Foucault M (1985a) The order of things. An archaeology of the human sciences. Tavistock, London

Foucault M (1985b) The archaeology of knowledge. Tavistock, London

Haaland C, van den Bosch CK (2015) Challenges and strategies for urban green-space planning in cities undergoing densification: a review. Urban For Urban Greening 14(4):760–771

Lapintie K (2005) The bird and the beast. In: Werquin et al (eds) Green structure and urban planning. COST Action C11 final report. Office for Official Publications of the European Communities, Luxembourg

Lapintie K, di Marino M (2019) Biodiversity, ecosystem services and environmental planning. In: Davoudi S, Cowell R, White I, Blanco H (eds) The Routledge companion to environmental planning. Routledge, Oxon/New York

Province of Milan (now Milan Metropolitan area) (2013) Tavola 4, Rete ecologica, Piano territoriale di coordinamento provinciale, approvazione 2013. http://allegati.provincia.milano.it/pianificazione/allegati/piano_territoriale/PTCP_vigente_2014/TAV4_rete_ecologica.pdf
Tiisala T (2015) Keeping it implicit: a defence of Foucault's archaeology of knowledge. J Am Philos Assoc 2015:653–673
Tyrväinen L, Ojala A, Korpela K, Lanki T, Tsunetsugu Y, Kagawa T (2014) The influence of urban green environments on stress relief measures: a field experiment. J Environ Psychol 38:1–9
Van den Berg AE, Jorgensen A, Wilson ER (2014) Evaluating restoration in urban green spaces: does setting type make a difference? Landsc Urban Plan 127:173–181
Wittgensein L (1961) Tractatus Logico-Philosophicus. Routledge & Kegan Paul, London

Chapter 5
The Project of the Green Infrastructure in Lombardy Region. A Resilient Spatial Structure for the Landscape Plan

Andrea Arcidiacono and Silvia Ronchi

Abstract This contribution describes the ongoing revision process of the Lombardy Region Landscape plan (PPR). As part of the landscape plan revision, the regional Ecosystem services (ES) mapping plays a key role in addressing strategies for biodiversity conservation and landscape protection. The aim is to update the plan's fact-finding contents and integrate landscape environmental, ecological and anthropic components and use a methodological approach to define plan contents, including spatial design elements and regulations. The design of the Green and Blue Infrastructure (GBI) and its different characterisation are supported and directly connected to the mapping and evaluation of ES, to preserve and protect landscape values and define strategies and planning actions to increase natural assets for human well-being in a resilient multi-scale perspective.

Keywords Landscape planning · Mapping ecosystem services · Habitat quality · Landscape regeneration · Multi-scalarity · Soil degradation · Green and Blue Infrastructure · Regional planning

5.1 Introduction

Lombardy is the Italian region with the largest surface area (23,864 sq.km) and population (more than 10 million). It is the most economically important, with a gross domestic product per capita of more than 37,000 Euro, higher than the Italian and European average, making up about 22% of the national total. The main 100 multinationals operating in Italy and more than 20 universities with more than 200 academic programmes are based in Lombardy. The Lombardy region has the

A. Arcidiacono (✉) · S. Ronchi
Department of Architecture and Urban Studies, Politecnico di Milano, Milano, Italy
e-mail: andrea.arcidiacono@polimi.it

A. Arcidiacono, S. Ronchi (eds.), *Ecosystem Services and Green Infrastructure*, Cities and Nature, https://doi.org/10.1007/978-3-030-54345-7_5

highest level of urbanisation making up more than 13% of the regional surface with 1507 municipalities (19% of the Italian total) and with an average population density of more than 420 inhabitants/sq.km. It has the highest land take and soil sealing rates. In 2018, only 633 hectares of land were anthropised (ISPRA 2019) in the Lombardy region. This is in line with the decrease of the national trend of urbanisation phenomena of agricultural and natural areas. However, the intensity of land take process recorded throughout the first decade of the 2000s was exceptional, with an average speed that exceeded 4000 hectares per year of land transformed (about 12 hectares per day), for more than 45,000 hectares urbanised between 1999 and 2012 (Masini et al. 2014) – an area corresponding to about 2.5 times that of a city as large as Milan (Arcidiacono et al. 2015a), with irreversible effects and impact on the availability of environmental and ecological resources, and on the conservation and appreciation of its landscape features (Arcidiacono et al. 2018a). Paradoxically, urban growth and land take have been accompanied by an increase of the decommissioning and degradation phenomena. More than 3400 areas covering more than 5000 hectares of regional surface suffered significant abandonment and underutilisation. They include large and medium industrial factories, traditionally located in central and peri-urban areas; but moreover, this trend increasingly involves different activities of the Lombardy region economy (medium and large commercial settlements, logistics, tertiary and tourism), infrastructures, public services and residential settlements. This is a phenomenon that deeply affects those parts of the Lombardy region, which suffer the greatest hardships. These are regional 'internal areas', affected by economic and demographic contraction, where the landscape's profound changes coincide with widespread degradation of natural elements and traditional cultural and identity values.

Although urban development and growth process continues to transform natural resources and threaten landscape values intensely (Arcidiacono et al. 2015a), and degradation and abandonment phenomena are increasingly affecting large parts of the region, the Lombardy region maintains exceptional qualities and variety of landscape features. About 40% of the region is mountainous, consisting of Alpine, pre-Alpine systems and a portion of the Apennine chain, with high natural and ecological value. Mountain landscape is an area of ecological uniqueness, a water reserve and a place under environmental protection, but on the other hand, it is mainly considered a source of natural resources that can be exploited. The systems of lakes and water bodies, with the three main basins Como, Maggiore and Garda Lakes and the large river systems (e.g. Po, Ticino, Adda, Oglio and Mincio) form the regional great landscape structure. The plain, which occupies about 13% of the regional surface, is the most productive in the Country, together with the neighbouring regions of Emilia-Romagna and Veneto, and it is rich in historical and cultural values, traditional settlements and small historical centres. The regional tradition and millenary agricultural structure defined the Lombardy plain landscape, together with the natural and artificial hydrographic system.

The region is under 'legal protection' for more than half of its area (Legislative Decree no. 42/2004 – Cultural Heritage Code) and about 22% consists of regional and national parks. This wide system of regulatory constraints has proved to be minimally effective in ensuring protection and appreciation of the landscape heritage against urbanisation threats and pressures, denoting its limitations in territorial protection and landscape management. Most of the protected areas do not have protection management measures and criteria; that is there are no guidelines and regulations for conservation or enhancement of areas with landscape value (Cazzani and Coloru 2016). This lack can lead to the loss of this heritage and requires an effort in uniting the heritage protection approach with an adequate landscape contextualisation and planning. What is needed is an integrated approach to landscape protection and management.

The Territorial Regional Plan (*Piano Territoriale Regionale* – PTR), approved in 2010, is the current planning tool which governs and manages the Lombardy territory and landscape. This is a complex and problematic territory that features significant environmental and landscape values and qualities. Under the regional legislative framework (Law no.12/2005), the PTR integrates and has the role also of a Landscape plan (*Piano Paesaggistico Regionale* – PPR). After about 5 years of PTR/PPR implementation, the plan's critical issues and substantial operational ineffectiveness in implementing regional objectives and strategies, particularly at local scale, were evident. In 2015, a Landscape Plan revision was initiated for this reason. This process involved the Department of Architecture and Urban Studies (DAStU)[1] of Politecnico di Milano, based on a partnership with the Lombardy Region Authority. The main objective was strengthening PPR effectiveness by influencing the landscape protection and planning tools at different scales and increasing the relevance of environmental and ecological components in the Plan's spatial design, defining planning strategies and regulations.

5.2 Integrate Complex Value Systems and Update Knowledge. Mapping and Evaluating Ecosystem Services to Support the Landscape Plan

The research studies and project proposal carried out by the DAStU working group rose from a need to address the many and articulated Lombardy landscapes, which used a meaningful approach that treated landscape planning in a non-sectorial

[1] Since 2015, the Department of Architecture and Urban Studies (DAStU) of Politecnico di Milano has been working in a partnership with the Lombardy Region to carry out research and studies to support the Regional Landscape Plan (PPR) revision. The regional department responsible for the planning process is part of the 'Regional and Civil Protection Directorate-General'. The coordination and scientific management is entrusted to Luisa Pedrazzini, who is the head of the 'Landscape Department', and operational coordination are entrusted to Cinzia Pedrotti. The DAStU scientific management and coordination is entrusted to Andrea Arcidiacono, while the protected areas manager is Alberta Cazzani. The research group is composed of Monica Aresi, Stefano Coloru, Paolo Dilda, Viviana di Martino, Marika Fior, Federico Ghirardelli, Daniela Giannoccaro, Giulio Giordano, Carlo Manfredi, Laura Pogliani, Silvia Ronchi, Silvia Restelli, Stefano Salata, and Francesco Secchi.

manner. This would make PPR an essential tool within a wider set of integrated projects and policies, in line with the European Landscape Convention (2000) reference framework.

Two crucial issues were addressed. The first was the problematisation of the nexus between landscape and territorial development, where the landscape is not only perceived from a conservation and aesthetic viewpoint. Instead, the landscape is conceived as a project approach for a more sustainable development perspective where it can support new ways of life and economies. The second issue regards the assumption of a deep continuity and intertwining between landscape, environment and ecology. This includes the production of new knowledge, frameworks and effective rules.

The landscape, as a concept in-between new development trajectories and deep rootedness within an extensive environmental and ecological issue, is the central piece of a territorial project, composed by protection rules and criteria and regeneration strategies for the resilient appreciation of heritage, resources and values. From this perspective, the landscape management is not only as mere protection of historical values but an active approach that has an impact, using actions and regulatory and planning tools, to enhance existing natural and cultural resources, regenerate degraded areas and limiting and mitigating possible threats.

Within this field of meaning, specific studies and research contributions have been carried out to strengthen the plan's environmental, ecological component, fact-finding and interpretative aspects of the Lombardy landscape structures and conservation and regeneration planning strategies. This aims at strengthening the effectiveness of planning and regulatory tools, acting selectively in simplification and integration of planning tools, introducing guidelines and coordinated criteria of landscape and protection system management to influence planning processes locally. PPR revision objectives included a different Landscape plan connotation, aimed at overcoming a purely 'inventory' approach to landscape. This was traditionally assumed for the conservation of protected areas and elements. The new landscape planning approach strengthens it as an active governance tool, able to expand and differentiate its planning component within a territorial-based vision of the landscape as a system of relationships and interactions between anthropic and natural elements that include ordinary areas, increasingly connoted by marginality and residual use. A planning approach that limits and compensates landscape anthropic transformation processes (i.e. land take and soil sealing) and addresses specific degraded landscape regeneration processes. Strengthening the analytic and interpretative system has been a priority to support a reconfiguration of the PPR regulatory and planning contents (Arcidiacono et al. 2016) aiming at identifying the natural and anthropic pattern, which are the landscape's structural components. It aims at restoring the landscape value systems' complexity, as a convoluted process of interactions and mutual adaptations between environmental and anthropic systems.

Mapping and evaluating different Ecosystem services (ES) has been fundamental to redefining the plan's knowledge content from an ecologically oriented perspective. In recent years, ES have gained more attention in the scientific literature and academic research that has increasingly highlighted the relevance of an approach based on mapping, classifying and evaluating ecosystem functions (de Groot et al. 2002; Costanza 2008; Fisher et al. 2009; Haase et al. 2014; Costanza et al. 2017) and possible applications to support decision-making processes for landscape planning and management (Burkhard et al. 2013; Hansen et al. 2015). Despite the growing interest in ES, not only from the scientific and academic community but also from policymakers, only few planning cases include ES in defining territorial development and landscape strategies (de Groot et al. 2010). It has been demonstrated that changes in land use and cover, and spatial planning, have a direct and often irreversible impact on ecosystems. The need to test and integrate ES evaluation and quantification in decision-making for landscape planning and management processes seems to be recognised and shared by many researchers (de Groot et al. 2002). However, it is necessary to develop an adequate ES definition and classification to integrate and guide decision-makers towards shared and sustainable choices to manage natural resources and landscape planning (Maes et al. 2012; Gomez-Baggethun and Barton 2013).

In the Lombardy Landscape Plan, mapping the 'Habitat quality' has led to the recognition of the landscape as an ecosystem network (Ferrari and Pezzi 2013; Salata et al. 2016). Interpreting environmental and ecological features of different landscapes has been combined with an analysis of the urban morphologies giving a spatial representation of critical issues and development forms (e.g. porosity, dispersion, fragmentation, compactness, urban shape) and their effects on ecosystem functions (Ronchi et al. 2018). In addition, the analysis of the agricultural landscape conducted considering the diversification of agricultural practices and the presence of natural and anthropic elements in agricultural patterns provides an evaluation of rural values.

In-depth studies on different landscape structures (natural, anthropic and rural) showed the connection between the natural and anthropic features of landscape ecosystems highlighting the relationships of balance and reciprocity between urban systems and agricultural and natural areas, and the related interference effects. A further analytical study aimed at identifying and assessing situations of pressure, degradation, risk and phenomena that threaten ecosystem functions with potential impacts on landscape values and recognisability (Fig. 5.1). These aspects were carried out by mapping land take processes, considering three thresholds (1954, 1999, 2012), integrated with an analysis of urban transformations forecasted by local urban plans in-force. In addition, there were identified existing landscape degradation and deterioration elements, including abandonment and decommissioning of production settlements (landfills, abandoned industrial areas, contaminated sites, abandoned quarries), large infrastructures, energy production plants, logistics, large shopping malls; contraction of agricultural production activities, with the related depopulation of highly fragile landscapes (e.g. mountains), and monoculture intensification, which causes increasingly widespread landscape standardisation. Phenomena due to human activities and to urban planning transformations; changes in land use that deeply threaten not only the landscape values but the capability of

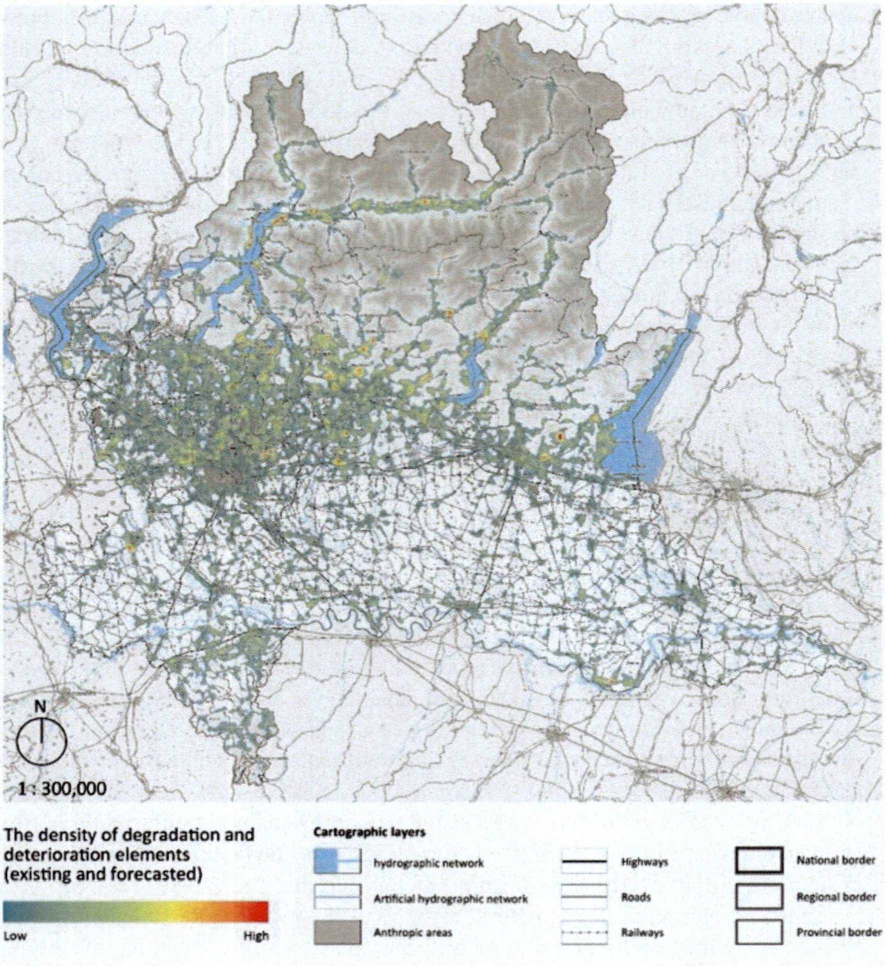

Fig. 5.1 The density of degradation and deterioration elements (existing and forecasted) (Authors elaboration derived by the Regional Landscape Plan, Map QC 6.1 *I territori d'attenzione*)

natural systems to provide ES and produce benefits for human well-being (Millennium Ecosystem Assessment 2005).

To address spatial planning through a more conscious approach to ecosystem functions and related services, the landscape scale is the most appropriate and effective for the definition of strategies to maintain and enhance ES (Albert et al. 2016). Landscape was defined by the European Convention (2000) as 'an area, as perceived by people, whose character is the result of the action and interaction of natural and/or human factors', and by landscape ecology as 'heterogeneous land area composed of a cluster of interacting ecosystems that is repeated in similar form throughout' (Forman and Godron 1986). Adopting an ES-based approach can play a key role in configuring planning tools to guide choices towards maintaining biodiversity and protecting the landscape for human well-being. The ES evaluation, associated with a mapping of territorial structures and landscape patterns, can directly identify strategies for landscape planning and define regulatory and planning guidelines for landscape protection and enhancement. Qualitative and quantitative ES evaluation models, which provide an analytical and reconnaissance framework of services provided by a landscape (Crossman et al. 2013), can affect operationally spatial plans and projects orienting guidelines and management criteria for its protection and safeguard. The ES evaluation tools and methods are useful to identify landscape areas with similar features and values. Based on these, it is possible to define specific guidelines and criteria for the landscape protection, recovery and improvement, combining concepts of landscape naturalness and ecology, protection and preservation, with traditional aesthetic-perceptual elements and values.

In the PPR experience, the approach and methodology based on the ecosystem quality analysis became the fact-finding reference on which to anchor the construction of the Plan's contents for two main themes (Arcidiacono et al. 2015b). The first refers to the identification of areas with similar 'highly natural' conditions that need strategies of conservation and appreciation of habitats. The second concerns the characterisation and the design of the Green Infrastructure (GI), as a resilient, multiscale landscape structure (McPhearson et al. 2015) to be used in planning.

The identification of areas with 'high habitat quality' was developed as an analytical basis through ES mapping using InVEST software (Integrated Valuation of Ecosystem Services and Tradeoffs)[2] that allows to assess the benefits provided by ecosystems and the possible trade-offs determined by alternative scenarios (Arcidiacono et al. 2015c). InVEST is an ES-based geographical, economic and ecological accounting tool for regional and urban planning in terms of restoring and conserving the soil's natural capacity to provide ecosystem services (Salata et al. 2017a).

The Habitat Quality indicator expresses (with values ranging from 0 to 1) overall ecological quality based on proximity of the habitat to human land uses and the degree of disturbance caused by them. It was used as a synthetic indicator and as a proxy of the ecological state of the Region (Salata et al. 2017b). The indicator allowed to re-shape the protected areas designated by the regional law and by the former Landscape Plan (Fig. 5.2).

[2]https://naturalcapitalproject.stanford.edu/software/invest

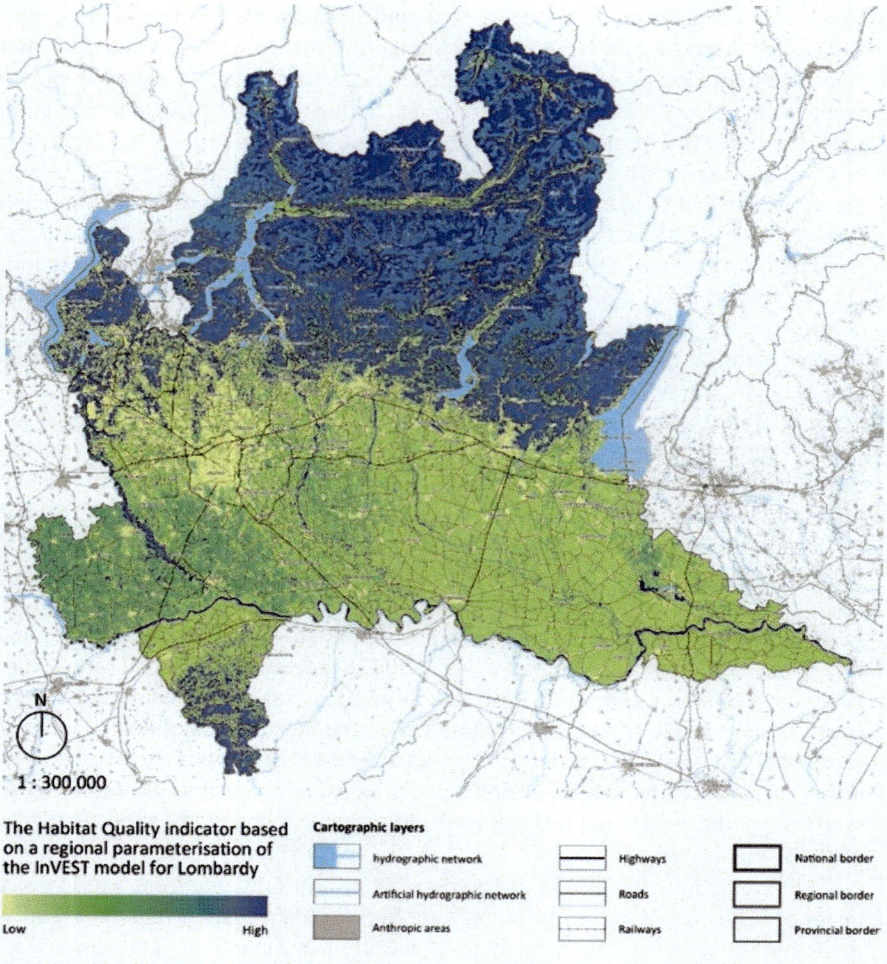

Fig. 5.2 The Habitat Quality indicator based on a regional parametrisation of the InVEST model for Lombardy (Authors elaboration derived by the Regional Landscape Plan, Map QC 3.1 *Habitat Quality*)

5.3 The Green Infrastructure (GI) as a Strategic Backbone of the Landscape Plan

The construction of the new regional landscape plan is directly connected to the deepening of the ES system knowledge and mapping. This directly affect the design of the spatial structure of the plan and the definition of the regulatory tools. The Green Infrastructure (GI) is the strategic design component of the plan, which goes beyond the meaning of an exclusively environmental and ecological network (Davies et al. 2006), becoming the regional landscape backbone ES-based

(Lafortezza et al. 2013). This combines elements of conservation of existing natural and historical-cultural values; measures to limit and protect landscape from increasing anthropic pressures (Maes et al. 2014); and a design approach aimed at the recovery and regeneration of degraded landscapes, to create new ecologically oriented landscapes and increase natural capital (European Commission 2013). A resilient and multi-scale landscape structure that allows the evaluation of the impacts related to forecasted urban transformations (also including infrastructures) set by planning and programming tools at different scales (Zulian et al. 2018). The GI spatial and functional definition started from mapping and evaluating ecosystem services considering three different landscape types:

1. Natural landscapes, based on the Habitat quality value (Fig. 5.2), that is a service mainly supporting the naturalistic-ecological functionality of the Lombardy area (Salata et al. 2017b)
2. Anthropic landscapes, featuring historical and cultural heritage sites (mainly derived by the Code of the Cultural and Landscape Heritage) derived from a density analysis of the historical-cultural values of Lombardy region's urban and suburban areas, and their connectivity provided by 'slow mobility' infrastructures (cycle and pedestrian paths)
3. Rural landscapes, based on the analysis of the agricultural patterns and 'mosaics' previously described, that is a provisioning and regulating services derived from mapping production, ecological (biodiversity of rural areas) and landscape value of land used for agricultural activities

These values were represented using maps showing the regional distribution of these three categories (Haase et al. 2014).

The various components (natural, agricultural or cultural-historical) have defined the regional GI structural features for which actions of landscape protection, regeneration and reconstruction are planned and articulated. They are governed by the normative component of the GI itself (Arcidiacono et al. 2018b). The GI, configured as a regional priority landscape infrastructure, becomes a tool for the protection of landscape values supporting measures to contain land take, and a spatial planning tool for landscape regeneration, capable of influencing and guiding urban planning process (Fig. 5.3).

The GI functional definition made it possible to outline objectives and measures related to the implementation of landscape projects based on multi-systemic components for areas included in the network design, in a trans-scalar perspective. The GI defines the regional landscape structures and applies a method to different scale – the intermediate territorial scale, within homogeneous geographical landscape areas, defined by the plan as territorial systems which are consistent from the landscape and ecosystem point of view; and the local scale addressed by the municipal urban planning. The proactive approach of the GI project limits its up-imposed character and opens it up to a shared local landscape planning that enhances its implementation effectiveness. To implement landscape projects using the Local Green Infrastructure, the contribution of regional landscape plan knowledge frameworks must be complemented by an analysis of soil values and the environmental

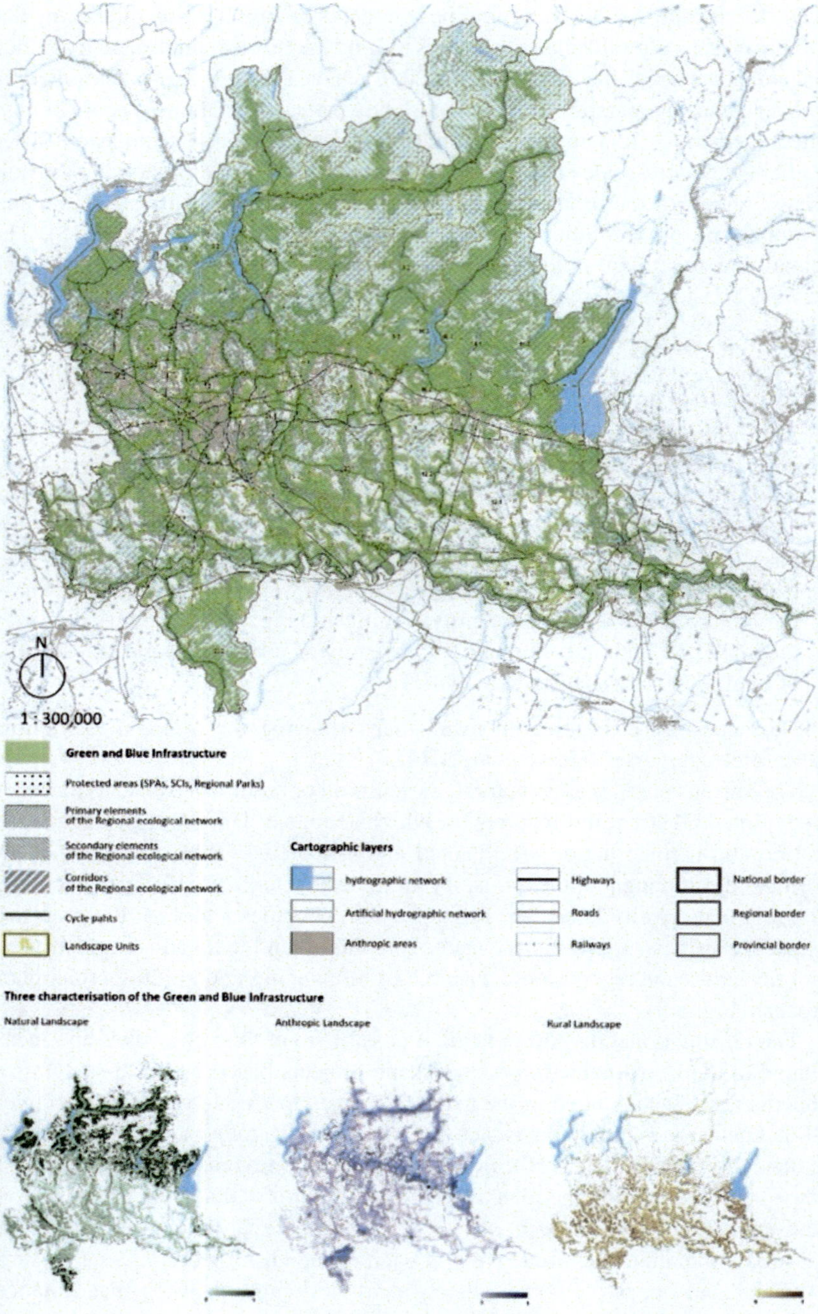

Fig. 5.3 Regional Green and Blue Infrastructure (Authors elaboration derived by the Regional Landscape Plan, Map PR 4.1 *Rete Verde Regionale*)

and ecosystem pressures withstood on the urban scale. The focus is on strengthening regulating ecosystem services, which are less relevant in the identification and programming at the regional scale and which find their suitability in local mapping by mitigating or compensating the environmental pressures exerted by land-use transformations. Regulating services primarily contribute to the well-being and quality of life of settled communities, and they protect from environmental risks related to the control of hydrological regimes, erosion, disruption and water and air quality.

5.4 Conclusions

The PPR revision process is an important opportunity for field testing in landscape planning, ecosystem evaluation methodologies and approaches.

The adoption of an ecosystem approach in a regional landscape plan is rare yet, because the ES concept is still not widely recognised and used outside the scientific discourse, and because there are few practical experiences of its integration in the planning processes at different scales.

ES evaluation can change the approach to sustainability and resiliency in planning and decision-making, there is still no rigorous, systematic and shared methodology for their integration into planning processes (Gret-Regamey et al. 2016). This underlines the need to increase ES awareness among policy-makers (Saarikoski et al. 2018) and citizens to spread awareness of ecosystem value in improving and enhancing human well-being (Dodge et al. 2012).

The case of Landscape plan revision showed how the results of an ecosystem analysis were functional for Green regional infrastructure design, 'translating' the ES concepts and contents into a multi-scale landscape project. The adoption of an ES-based approach in a regional landscape plan implied the need to experiment with methods and tools to scale down PPR contents in provincial, supra-local and local planning and programming tools. This is undoubtedly a crucial and problematic issue for the dissemination of awareness and knowledge on ES, intending further to integrate it into spatial planning at different scales. Applying an ES-based approach directly affects plan's regulatory and design provisions including conservation of highly natural areas and the definition of strategies and active measures for landscape regeneration and value improvement.

References

Albert C, Galler C, Hermes J, Neuendorf F, Haaren C, Von Lovett A (2016) Applying ecosystem services indicators in landscape planning and management: the ES-in-Planning framework. Ecol Indic 61:100–113. https://doi.org/10.1016/j.ecolind.2015.03.029

Arcidiacono A, Ronchi S, Salata S (2015a) Ancora consumo di suolo in Lombardia? Previsioni di trasformazione nei Piani di Governo del Territorio. In: Arcidiacono A, Di Simine D, Oliva F, Salata S, Ronchi S (eds) Nuove sfide per il suolo. Rapporto 2016 CRCS. INU Edizioni, Roma

Arcidiacono A, Ronchi S, Salata S (2015b) L'approccio ecosistemico nella pianificazione di area vasta: il Piano Paesaggistico Regionale della Lombardia. In: Arcidiacono A, Di Simine D, Oliva F, Salata S, Ronchi S (eds) Nuove sfide per il suolo. Rapporto 2016, Centro di Ricerca sui Consumi di Suolo CRCS. INU Edizioni, Roma

Arcidiacono A, Ronchi S, Salata S (2015c) Ecosystem services assessment using InVEST as a tool to support decision making process: critical issues and opportunities. In: Computational science and its applications-ICCSA 2015. Springer International Publishing, Cham, pp 35–49

Arcidiacono A, Fior M, Pasqui G (2016) Governare il paesaggio. Sfide per il PPR lombardo. Territorio 77:15–24

Arcidiacono A, Ronchi S, Salata S (2018a) I processi di variazione d'uso del suolo avvenuti tra il 1999 e il 2015 nelle aree di interesse paesaggistico della Lombardia. In: Arcidiacono A, Di Simine D, Ronchi S, Salata S (eds) Consumo di suolo, servizi ecosistemici e green infrastructures: Caratteri territoriali, approcci disciplinari e progetti innovativi. Rapporto 2018 CRCS. INU Edizioni, Roma

Arcidiacono A, Salata S, Ronchi S (2018b) Un approccio ecosistemico al progetto delle infrastrutture verdi nella pianificazione paesaggistica. Urbanistica 159:102–114

Burkhard B, Crossman N, Nedkov S et al (2013) Mapping and modelling ecosystem services for science, policy and practice. Ecosyst Serv 4:1–3. https://doi.org/10.1016/j.ecoser.2013.04.005

Cazzani A, Coloru S (2016) Criteri per la gestione attiva dei vincoli e degli ambiti di tutela paesaggistica in Lombardia. Territorio 77: 31–37, FrancoAngeli

Convenzione Europea del Paesaggio (2000) Firenze

Costanza R (2008) Ecosystem services: multiple classification systems are needed. Biol Conserv 141:350–352. https://doi.org/10.1016/j.biocon.2007.12.020

Costanza R, de Groot R, Braat L et al (2017) Twenty years of ecosystem services: how far have we come and how far do we still need to go? Ecosyst Serv 28:1–16. https://doi.org/10.1016/j.ecoser.2017.09.008

Crossman ND, Burkhard B, Nedkov S et al (2013) A blueprint for mapping and modelling ecosystem services. Ecosyst Serv 4:4–14. https://doi.org/10.1016/j.ecoser.2013.02.001

Davies C, MacFarlane R, McGloin C, Roe M (2006) Green infrastructure planning guide, p 45. [online] URL: http://www.scribd.com/doc/

de Groot R, Wilson MA, Boumans RMJ (2002) A typology for the classification, description and valuation of ecosystem functions, goods and services. Ecol Econ 41:393–408. https://doi.org/10.1016/S0921-8009(02)00089-7

de Groot R, Alkemade R, Braat L et al (2010) Challenges in integrating the concept of ecosystem services and values in landscape planning, management and decision making. Ecol Complex 7:260–272. https://doi.org/10.1016/j.ecocom.2009.10.006

Dodge R, Daly A, Huyton J, Sanders L (2012) The challenge of defining wellbeing. Int J Wellbeing 2:222–235. https://doi.org/10.5502/ijw.v2i3.4

European Commission (2013) Communication from the Commission to the European Parliament, the Council, the European economic and social Committee and the Committee of the regions. Green Infrastructure (GI) — enhancing Europe's Natural Capital {SWD(2013) 155 final}. European Commission, Bruxelles

Ferrari C, Pezzi G (2013) L'ecologia del paesaggio. Il Mulino, Bologna

Fisher B, Turner RK, Morling P (2009) Defining and classifying ecosystem services for decision making. Ecol Econ 68:643–653. https://doi.org/10.1016/j.ecolecon.2008.09.014

Forman RTT, Godron M (1986) Landscape ecology. John Wiley and Sons Ltd., New York

Gómez-Baggethun E, Barton DN (2013) Classifying and valuing ecosystem services for urban planning. Ecol Econ 86:235–245. https://doi.org/10.1016/j.ecolecon.2012.08.019

Grêt-Regamey A, Sirén E, Brunner SH, Weibel B (2016) Review of decision support tools to operationalize the ecosystem services concept. Ecosyst Serv April:0–1. https://doi.org/10.1016/j.ecoser.2016.10.012

Haase D, Larondelle N, Andersson E, Artmann M et al (2014) A quantitative review of urban ecosystem service assessments: concepts, models, and implementation. Ambio 43:413–433. https://doi.org/10.1007/s13280-014-0504-0

Hansen R, Frantzeskaki N, McPhearson T, Rall E et al (2015) The uptake of the ecosystem services concept in planning discourses of European and American cities. Ecosyst Serv 12:228–246. https://doi.org/10.1016/j.ecoser.2014.11.013

ISPRA (2019) Consumo di suolo, dinamiche territoriali e servizi ecosistemici, Rapporto 2019. ISPRA, Roma

Lafortezza R, Davies C, Sanesi G, Konijnendijk CCC (2013) Green Infrastructure as a tool to support spatial planning in European urban regions. iForest – Biogeosci For 6:102–108. https://doi.org/10.3832/ifor0723-006

Maes J, Egoh B, Willemen L et al (2012) Mapping ecosystem services for policy support and decision making in the European Union. Ecosyst Serv 1:31–39. https://doi.org/10.1016/j.ecoser.2012.06.004

Maes J, Barbosa A, Baranzelli C et al (2014) More green infrastructure is required to maintain ecosystem services under current trends in land-use change in Europe. Landsc Ecol 30:517–534. https://doi.org/10.1007/s10980-014-0083-2

Masini M, Montagnana S, Fiorini M (2014) Evoluzione dell'uso del suolo dal 1999 al 2012. Studi e Ricerche, Regione Lombardia

McPhearson T, Andersson E, Elmqvist T, Frantzeskaki N (2015) Resilience of and through urban ecosystem services. Ecosyst Serv 12:152–156, Elsevier. https://doi.org/10.1016/j.ecoser.2014.07.012

Millenium Ecosystem Assessment (2005) Ecosystems and human well-being: our human planet. Island Press, ISBN:9781559633864

Ronchi S, Salata S, Arcidiacono A (2018) An indicator of urban morphology for landscape planning in Lombardy (Italy). Manag Environ Qual Int J 29:623–642. https://doi.org/10.1108/MEQ-05-2017-0048

Saarikoski H, Primmer E, Saarela SR et al (2018) Institutional challenges in putting ecosystem service knowledge in practice. Ecosyst Serv 29:579–598. https://doi.org/10.1016/j.ecoser.2017.07.019

Salata S, Ronchi S, Ghirardelli F (2016) I servizi ecosistemici a supporto della pianificazione paesaggistica. Territorio 77:45–52

Salata S, Garnero G, Barbieri C, Giaimo C (2017a) The integration of ecosystem services in planning: an evaluation of the nutrient retention model using InVEST software. Land 6:48. https://doi.org/10.3390/land6030048

Salata S, Ronchi S, Arcidiacono A, Ghirardelli F (2017b) Mapping habitat quality in the Lombardy region, Italy. One Ecosystem 2:e11402. https://doi.org/10.3897/oneeco.2.e11402

Zulian G, Thijssen M, Maes J (2018) Enhancing resilience of urban ecosystems through green infrastructure (EnRoute) progress report

Chapter 6
A Green Infrastructure in the Guidelines to Limit Land Consumption of the Friuli Venezia Giulia Regional Landscape Plan

Elisabetta Peccol, Mirko Pellegrini, and Mauro Pascolini

Abstract Land consumption still represents a real threat to the landscape of some Italian regions, like Friuli Venezia Giulia (FVG), that are without specific national and regional laws to limit land take. Regional landscape plans are essential planning tools for landscape protection and management and, among their objectives, pursue the enhancement of landscape quality and the containment of land consumption, also in areas with no outstanding landscape. In this context, the strategic component of the Regional Landscape Plan of FVG, among its various planning documents, provides for the drawing up of Guidelines to Limit Land Consumption (GLLC). This non-statutory planning document, embraces the concept of green infrastructure to promote a proactive approach for limiting land take. Hence, the GLLC envisages a Green Infrastructure Strategic Framework (GISF) that encompasses the entire regional territory and aims at providing a reference frame where green infrastructure can be identified and prioritized with targeted strategies. The GISF, once approved, is intended to support both the implementation of the regional landscape plan and of spatial and sector plans at different territorial levels, with particular regard for municipal master plans.

Keywords Green infrastructure framework · Green network · Open space multifunctionality · Land take · Urban–rural relationships · Friuli Venezia Giulia Regional Landscape Plan

E. Peccol (✉)
Department of Agricultural, Food, Environmental and Animal Sciences (DI4A), University of Udine, Udine, Italy
e-mail: elisabetta.peccol@uniud.it

M. Pellegrini · M. Pascolini
Department of Languages and Literatures, Communication, Education and Society (DILL), University of Udine, Udine, Italy

A. Arcidiacono, S. Ronchi (eds.), *Ecosystem Services and Green Infrastructure*, Cities and Nature, https://doi.org/10.1007/978-3-030-54345-7_6

6.1 Introduction

The growth of urban areas is a major threat to landscape preservation in Italy (Fiorini et al. 2019; Amato et al. 2017), due to consumption of natural and agricultural land and loss of both ecological and cultural landscape features.

The Legislative Decree no. 42/2004 "Code of the Cultural and Landscape Heritage" (CCLH) is the current law for the protection of cultural heritage and landscape assets that requires regional authorities to draw up regional landscape plans (RLP) as fundamental tools for landscape conservation and management.

The RLP, in general, analyzes local landscape patterns and identifies landscape units, characterized by different landscape policies, depending on the presence and quality of local landscape features. In particular, the RLP also defines transformations for each landscape unit that are compatible with the respective landscape values and sets specific objectives of limiting land take and enhancing landscape quality in the whole territory, including areas with no outstanding landscape. The RLP is a particularly important tool, since it targets all other plans (spatial, urban, sector, etc.) at different scales, up to municipal master plans.

The Regional Landscape Plan of Friuli Venezia Giulia (FVGRLP) was launched in 2014 and drawn up by the Landscape Protection and Biodiversity Service of the Regional Authority jointly with the Ministry for Cultural Heritage and Activities (MCHA). The plan is structured in a statutory component and a strategic component, integrated by a final management component.

The FVGRLP strategic part sets out the strategic objectives of the plan, and the framework for future projects aimed at enhancing landscape value and promoting sustainable local development, through a variety of planning tools and processes of different duration. In particular, it sets out unitary and functional policies aimed at guiding actions of protection, enhancement, reclamation, and upgrading of the Friuli Venezia Giulia (FVG) landscapes, which should ultimately contribute to the implementation of the plan.

The structure of the strategic part is, therefore, quite complex and includes, among others, the three Networks – Ecological, Slow Mobility, and Heritage Assets Networks – and four Guidelines. The latter, are planning tools on issues relevant to the FVG territory and are aimed at promoting good practice in territorial and landscape projects, especially when conducted at the local planning level. In this context, the Guidelines to Limit Land Consumption were drawn up by a research group,[1] within a research agreement between the University of Udine and the Regional Authority for providing scientific advice on the strategic part of the FVGPLP.

The FVGRLP came into force in May 2018, and the process of adaptation by the municipal masterplans is still under way. Besides, formal approval of the plan by the regional council concerned only part of the strategic planning documents, while the

[1] The research group of the Guidelines to Limit Land Consumption is composed by E. Peccol, M. Pellegrini, L. Cadez, L. Di Giusto, V. Ferrario (IUAV University), M. Pascolini.

Guidelines approval was postponed to a later date,[2] defined in a joint agreement by the Regional Authority and the MCHA.

6.2 The Guidelines to Limit Land Consumption

Friuli Venezia Giulia (FVG) is in the furthest Northeast of Italy and borders with Slovenia and Austria (Fig. 6.1). The region is known for its landscape diversity, yet artificial areas have grown disproportionally (62,300 hectares, equal to 7.9% of the regional territory, according to Corine Land cover 2018), especially on lowland

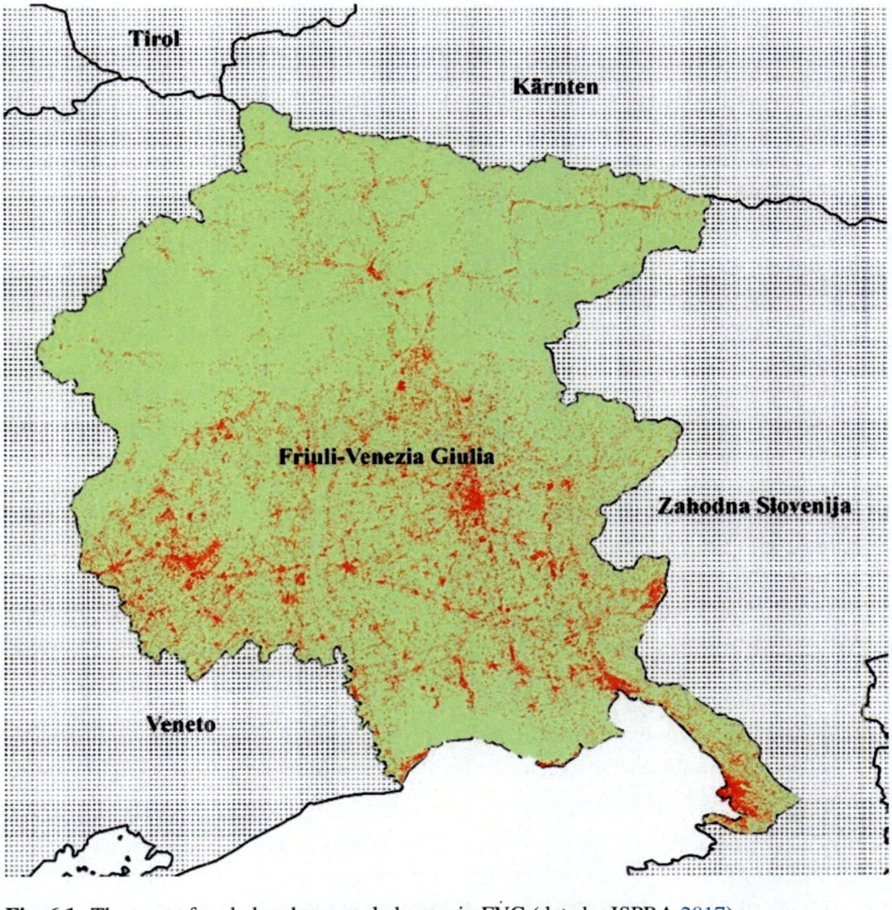

Fig. 6.1 The map of sealed and non-sealed areas in FVG (data by ISPRA 2017)

[2]At present, all FVGRLP guidelines are still awaiting approval, since in the meantime, there has been a change in the political governance of the region.

farmland. Moreover, urban growth has occurred as a phenomenon independent on demographic trends, having a sealed surface/per-capita in 2016 of 573 m^2 versus the Italian average of 380 m^2 (ISPRA 2017). Consequently, the rural landscape has undergone transformations, deriving principally from the constant encroachment into the countryside by productive and residential areas and the abandonment of agriculture, especially in the most disadvantaged areas like mountains.

The Guidelines to Limit Land Consumption (GLLC) is a nonstatutory planning tool of the FVGRLP, that embraces its principles and contributes to implementing the strategic objectives of the plan for the whole regional territory. In particular, general objective (GO) 3 "Halt the loss of biodiversity and ecosystem services" and GO 4 "Zero land take" are the most relevant for the GLLC. The GO 4 transposes the target set by the "Roadmap to a Resource Efficient Europe" (EC 2011) of "No Net Land Take by 2050" and lays down the following specific objectives (SO):

1. SO 4.1 Promote a sustainable use of common goods.
2. SO 4.2 Direct local planning towards preventing the loss of agricultural land.
3. SO 4.3 Pursue the strategy "To plan and to build on derelict urban areas".
4. SO 4.4 Pursue the conservation of open spaces and natural areas that provide carbon sequestration or other ecosystem services.

The GLLC has been structured in an analytical part, which provides an overview of the different types of land take in FVG, followed by a section that sets up a strategic framework for green infrastructure to support strategies to limit land consumption. Finally, in a third part, the GLLC defines a system for measuring and monitoring land take.

In this paper, we present the Green Infrastructure Strategic Framework (GISF) of the GLLC, with respect to the fundamental theme of the role of networked open spaces in planning the territory and landscapes of the FVG region.

Indeed, in order to limit land consumption, the GLLC does not rely on a binding approach, based on spatial constraints but on a GISF that places open space at the core of planning and decision-making across all fields – particularly in landscape management and spatial planning – and at all planning scales. The guidelines recognize open spaces as a strategic asset for regenerating territories, social well-being, and creating new opportunities for economic development. Indeed, if properly maintained and enhanced through appropriate planning and design, open spaces can provide benefits – ecosystem services – and vital resources that go beyond the landscape and environmental aspects and involve the social and economic dimension.

The GLLC main purpose is to contribute to implementing the FVGRLP strategic objectives. Hence, the GLLC primary objectives are to:

- Go beyond a constraint-based approach to the protection of open spaces, by setting up a GISF aimed at promoting virtuous processes – at various planning levels and involving multiple public and private actors – to enhance the potentials of open spaces for producing ecosystem services. Indeed, the GISF should be the framework for taking decisions and planning future land-use changes and the backbone for policies of urban and territorial regeneration, restoration of

degraded landscapes, ecological conservation and restoration, resilience building, agriculture strengthening and rural development.
- Ensure full recognition and leverage of the multifunctional benefits of green infrastructure among all relevant stakeholders in land-use change and landscape decisions (public agencies, planners, developers, etc.) and engage them in green infrastructure.
- Ensure a shared vision and a coordinated approach to green infrastructure across the FVG region and regional and national borders.
- Identify potential components of green infrastructure for small- and large-scale projects in FVG.
- Strengthen the strategies of the FVGRLP Ecological, Slow Mobility, and Heritage Assets Networks.
- Improve knowledge about the potential multibenefits from open spaces among citizens and local communities.
- Define a set of strategies for relevant issues for the FVG region in the matter of landscape and open space preservation and enhancement focusing on:

1. Awareness raising about green infrastructure and its benefits.
2. Preservation or re-use of open spaces in peri-urban areas.
3. Landscape mitigation for industrial and commercial areas.
4. Urban–rural connection and landscape practicability/accessibility.
5. Promoting agriculture (urban, peri-urban, and rural).

The GISF refers to the planning principles outlined in the next section.

6.3 From Land Consumption to a Green Infrastructure Strategic Framework

The Green Infrastructure Strategic Framework of the Guidelines to Limit Land Consumption of the FVG Regional Landscape Plan endorses a network of urban and rural open spaces[3] (GI), extended to the whole regional territory, that – if properly planned – holds present or potential capacity to deliver one or more ecosystem services. In planning processes, it must be treated as structural element of the regional territory, which is planned, implemented and managed as a multifunctional resource to sustain natural systems, to protect and enhance the regional landscape, to counteract land consumption and to improve life quality in urban and rural areas.

The ecological network (EN) of the FVGRLP, including natural protected areas, should be the backbone of the GI, while establishing structural and functional relations with it. Nevertheless, while the EN's main purpose is the conservation of

[3] In this context, the "open space" concept is intended as any open piece of land, in the urban, peri-urban, or rural context, that is undeveloped and permeable to water; open spaces are generally green and can be public or private, natural, or anthropic.

biodiversity and support of ecological processes, the GI's major purpose is to boost the delivery of ecosystem and landscape services at the different territorial levels (Termorshuizen and Opdam 2009).

In this sense, the GISF goes beyond the EN objectives, as it places GI as a fundamental planning framework, that takes open spaces as its core, including those ones in urban and peri-urban areas, without particular values and intensely transformed by human intervention.

The GISF theoretical framework embraces acknowledged GI planning principles (Benedict and McMahon 2002; Artmann et al. 2017; Hansen and Pauleit 2014) that are:

Multiscale approach: A multiscalar and hierarchical GI, implemented at different territorial levels: regional, district, municipal, and neighborhood.

Multiobject approach: GI comprises different interconnected components depending on scale, location across the rural–urban gradient, and planned function. Essential components, private or public ones, of GI as intended in the GISF are: the EN, the blue network, linear features (e.g. hedgerows and tree rows), agricultural areas, unimplemented urban land and unsealed reclaimable brownfields.

Connectivity: The GISF supports structural and functional connectivity to increase functionality of single open spaces and improve the performances of the network. The GISF also aims at promoting "social connectivity" and "social inclusion," by improving the offer of and accessibility to quality landscape and open spaces in both urban and rural areas.

Multifunctionality: The GISF promotes multifunctionality of open spaces, envisioned as "the combination of functions fulfilled by the GI which cover social, economic, ecological and cultural aspects" (Grădinaru and Hersperger 2019).

Integration: The GISF seeks the physical and functional integration of GI into gray infrastructure (e.g., transport), slow mobility networks, water management systems, and built-up structures.

6.4 Relevant Strategies of the Green Infrastructure Strategic Framework

6.4.1 Raise Awareness About Green Infrastructure and Its Benefits

At present, there is lack of a framework to establish GI at regional or local levels in FVG, and the concept of GI is seldom included in plans and policies or is treated as ecological network. Accordingly, the GISF recognizes the need to improve awareness, education, and capacity building about GI and its potential flow of benefits. In this context, some measures proposed are:

- To promote educational workshops about GI and its benefits in schools.
- To enable and support farmers, mainly through professional associations, to actively join GI projects by also committing their land.
- To support nonprofit associations and other stakeholders involved in environmental and landscape management, to promote information and dissemination about GI.
- To draw up technical guidelines by the Regional Authority to promote integration of GI into planning, especially at local territorial level.

6.4.2 Preserve or Re-use Open Spaces in Peri-urban Areas

Peri-urban space and its complex pattern of interlinked urban, rural, and "natural" areas result from urban growth, often with loss of landscape character and consequent degradation. Indeed, the encroachment of built-up areas on rural land has often formed enclosed and unused open spaces that are seldom suitable for productive agriculture. In these areas, residual unsealed open spaces and agricultural areas can play a major role in policies for limiting land consumption. In fact, it is essential to preserve residual open spaces, particularly those surrounded by built-up areas, as they can potentially perform different functions and supply provisioning (e.g., food production with agriculture), environmental or cultural (e.g., recreational) services, of a private or public nature (Viganò 2012). Strategies for peri-urban areas should principally be addressed in planning tools at different levels and, preferably, coordinated within large-scale structural planning strategies at municipal level. Accordingly, the measures for peri-urban areas concern:

- The preservation and enhancement of open spaces, particularly when unused or enclosed in urban fabric, with the establishment of new public green areas and collective facilities.
- The prevention of neighboring towns from merging into one another.
- The re-naturalization of derelict urban spaces, when present in natural contexts, with ecological/environmental rebalancing areas (woodlands, meadows, etc.).
- The prevention of further consumption of agricultural land by energy infrastructures and photovoltaic fields.
- The conservation and restoration of historical rural landscape structure (landform, land arrangement, path and trail networks, tree rows and hedgerows, historical and cultural features, historical agricultural practices, hydrographic network, etc.).

6.4.3 Landscape Mitigation for Industrial and Commercial Areas

Another crucial issue that affects landscape in FVG is land consumption by newly built commercial districts and industrial parks and the development of "commercial streets." In the past, these areas have been the drivers of high land consumption rates and consequent soil sealing for the many critical aspects involved in their development, such as large dimensions and the location along historic roads (e.g., the historic road S.S. 13 Pontebbana).

Indeed, in FVG, the surface per 1000 inhabitants of large retail chains is almost double the national average (687 m^2 vs 372 m^2 in 2015, data from Istituto di Ricerche Economiche e Sociali of FVG). These areas, for their mono-functionality and, in most cases, poor design quality, have produced landscape homologation due to the serial nature of buildings, lack of integration with the surrounding landscape, and poor relationships with the neighboring urban centers (Zanfi 2012). In this context, the GISF aims at:

- Encouraging practices for the regeneration and reuse of derelict productive, commercial, or logistic areas and buildings, and connecting them to GI.
- Limiting soil sealing of large open spaces (e.g., parking lots) with technical mitigation measures to preserve permeability.
- Mitigating the visual impact of these areas with natural features in order to achieve a better integration with the surrounding landscape.
- Avoiding mono-functionality of commercial settlements (shopping centers or streets) and establishing connections with neighboring towns through slow mobility infrastructures and GI.
- Encouraging energy saving and other green-economy best practices, with the re-use and regeneration of large areas and buildings, photovoltaic roofing systems, green roofs, and vegetable gardens.

6.4.4 Urban–Rural Connection and Landscape Practicability/Accessibility

The GISF promotes urban–rural connection, intended as a green multifunctional network in a system of widespread open spaces and corridors (greenway and parkway) integrated with slow mobility networks connecting green areas, from urban centers and the open countryside, across peri-urban areas. The result of this approach is a new idea of territorial space, in which the city and the countryside, are structurally and functionally related without "clear boundaries"(Secchi and Viganò 2011).

Through a soft mobility system of bicycle lanes, footpaths, and tourist and recreation routes, access to the landscape can be increased. Measures for pursuing an urban–rural connection should aim at:

- Restoring the physical and functional relationship between urban and rural areas, through the connection between natural or rural spaces and urban spaces (public and otherwise), in order to reconnect urban areas with the most valuable landscapes.
- Developing or strengthening the FVGRLP "Slow Mobility Network" between urban areas and the rural landscape, though allowing citizens and also tourists, to move along greenways or parkways in both directions without barriers.
- Recovering decommissioned and unused railways to convert them into greenways.

6.4.5 *Promote Urban, Peri-urban, and Rural Agriculture*

Farmland – especially if low-intensity – and other open spaces potentially suitable for agriculture, are regarded by the GISF as fundamental components of GI for the role they can play for delivering landscape and ecosystem services key for society and local community well-being.

Indeed, there is increasing recognition of the importance of agriculture and its spaces not only for supplying provisioning services, but also for developing strategies for landscape restoration and management, urban resilience (e.g., food and climate), slow mobility, recreation and leisure, and public health (Rolf et al. 2018; Balzan et al. 2018; Brinkley 2012; Rolf et al. 2018; Gullino et al. 2018). In particular, multifunctional agriculture and urban agriculture are gaining international attention for the benefits they can provide, particularly in urban and peri-urban areas such as: renewing and regenerating vacant spaces, improving biodiversity, mitigating climate change effects, food security, etc. (Rolf et al. 2018; Artmann and Sartison 2018; Gullino et al. 2018; Rogers and Hiner 2016; Sarker et al. 2019).

Hence, the GISF devotes special attention to these forms of agriculture. In particular, it recognizes the need to support and promote agriculture in plans and programs at all territorial levels and, to coordinate planning and agricultural policies (e.g., rural development programs).

In this context, we have selected some objectives and measures endorsed in the GISF as follows.

6.4.5.1 Objective of Protecting and Reusing Agricultural Land

- Limit new development on agricultural areas or other open spaces potentially suitable for agriculture, in accordance with the FVGRLP strategic objective of "Zero land take".
- Gain land for agriculture, by converting, in municipal master plans, selected undeveloped urban areas back to agricultural land use.
- Establish a "Bank of agricultural land" for the FVG region (e.g., by the Regional Authority), intended as a dynamic digital georeferenced inventory of abandoned

or unused land suitable for agricultural use and made available to third parties (especially young farmers) by renting or granting.

6.4.5.2 Objective of Promoting Multifunctional Agriculture

Agriculture multifunctionality entails the joint production of commodities and non-commodities by the agricultural sector (e.g., agro-tourism, social and care services, environmental services) and, if properly managed, may provide multiple landscape services, particularly in peri-urban areas (Zasada 2011). The GISF endorses the promotion of farm level multifunctionality within projects and initiatives aimed at coordinating the provision of landscape services at a wider territorial scale (Gullino et al. 2018; Rolf et al. 2018). In the GISF, the promotion of multifunctional agriculture relies on the following:

- Supporting farming practices that sustain biodiversity, soil and water protection, visual amenities, such as extensive grassland management or organic farming
- Shortening the food chain by supporting direct sale methods (e.g., farmers markets in urban areas)
- Promoting and supporting the establishment of social and educational farming, particularly in peri-urban areas
- Involving and coordinating multifunctional farms into networked initiatives for landscape management and enhancement (e.g., measures of RDPs)
- Promoting initiatives to boost local food products, especially in farms located along tourist and cultural routes
- Establishing "agricultural parks" for territorial development and rural landscape enhancement, as a planning tool to protect and create conditions for promoting and managing multifunctional agricultural land

6.4.5.3 Objective of Promoting Urban Agriculture

The GISF recognizes and promotes all forms of urban agriculture (UA)[4] (e.g., allotments and private gardens, rooftop gardens, community, educational, and therapeutic gardens) and for this purpose makes provisions for:

- Establishing a UA network involving all relevant stakeholders for developing UA projects and encouraging citizens' engagement and more inclusive and participatory local decision-making
- Developing a georeferenced inventory of spaces potentially suitable for prioritizing UA projects at municipal level

[4] Urban agriculture is meant as "all actors, communities, activities, places and economies that focus on biological production in a spatial context, which – according to local standards – is categorized as 'urban'" (Lohrberg et al. 2016).

- Planning spaces for UA as part of a wider network of open spaces, integrated with the FVGRLP Slow Mobility Network
- Designing places of care, retirement homes, schools, or disadvantaged city boroughs integrated with spaces where UA can unlock its social, educational, and public health benefits
- Promoting rooftop agriculture and urban gardening, also in temporary forms, within the planning regulatory framework.

6.5 Conclusions

The strategic component of the Regional Landscape Plan of Friuli Venezia Giulia makes provision for the drawing up of Guidelines to Limit Land Consumption on the whole regional territory. Indeed, Friuli Venezia Giulia is one of the Italian regions with the highest per capita land consumption, and, hence, a widespread demand for the containment of land consumption has grown among the population over the years. This issue also clearly emerged during the participatory process of the FVGRLP (RAFVGSPB 2018), where soil sealing, large derelict areas (mainly military), and brownfields were highlighted as major problems by the participants in the process.

The GLLC, while including a section required by the Regional Authority for monitoring and measuring land consumption, in its core section promotes a Green Infrastructure Strategic Framework that places open spaces as a crucial asset for targeting the strategic objectives of the FVGRLP, and also for pursuing urban revitalization and regeneration, by improving social well-being and creating new opportunities for green economic development.

There is not, at present, a framework to establish GI at regional or local level in FVG, and the concept of GI is seldom included in plans and policies or it is regarded as ecological network. The implementation of the FVGRLP Strategic Ecological Network is fundamental for biodiversity and ecological connectivity and could act as the backbone of future GI in the FVG region, yet GI goes further since can act as a multifunctional device capable of releasing its benefits over the whole regional territory.

The GISF of the GLLC is a first attempt to integrate the concept of GI into the FVG planning system, as already occurred in other European and non-European countries. In particular, the GISF is distinctive with respect to the concept of GI as intended in the COM 249/2013 final "Green Infrastructure (GI) – Enhancing Europe's Natural Capital" (EC 2013), since it emphasizes the crucial role that spaces for agriculture, as part of GI, may play for landscape conservation and management and for delivering social and environmental benefits in peri-urban and urban areas.

In the GLLC, we proposed a system of strategies and guidelines aimed at increasing the "quality" of planning and, accordingly, of the landscape. The purpose is to set a framework to define a structural network of open spaces able to renew, relink, or establish some "relationship" between different territorial components (urban, peri-urban, rural, residual, etc.). A fundamental aspect for the integration of GI in planning practices is the role research, and landscape planning can play in guaranteeing high-quality design for open spaces.

In this context and with the growing and widespread interest of public opinion in the consequences of climate change and impacts of soil sealing, it is essential to entirely approve and implement the FVGRLP, including the strategic part in its whole, as originally designed. In this context, the GISF could act as an effective framework to implement active policies aimed at limiting land consumption and building a widespread network of green open spaces at various planning levels in the regional context.

References

Amato F, Martellozzo F, Nolè G, Murgante B (2017) Preserving cultural heritage by supporting landscape planning with quantitative predictions of soil consumption. J Cult Herit 23:44–54

Artmann M, Sartison K (2018) The role of urban agriculture as a nature-based solution: a review for developing a systemic assessment framework. Sustainability 10:1937; Available via DIALOG. https://www.mdpi.com/2071-1050/10/6/1937. Accessed 27 Aug 2019

Artmann M, Bastian O, Grunewald K (2017) Using the concepts of green infrastructure and ecosystem services to specify Leitbilder for compact and green cities—the example of the landscape plan of Dresden (Germany). Sustainability 9:198. Available via DIALOG. https://www.mdpi.com/2071-1050/9/2/198. Accessed 27 Aug 2019

Balzan MV, Caruana J, Zammit A (2018) Assessing the capacity and flow of ecosystem services in multifunctional landscapes: evidence of a rural-urban gradient in a Mediterranean small island state. Land Use Policy 75:711–725

Benedict MD, McMahon ED (2002) Green infrastructure: smart conservation for the 21st century. Renewable Resourc J Autumn Edition:12–17

Brinkley C (2012) Evaluating the benefits of peri-urban agriculture. J Plan Lit 27(3):259–269

European Commission (EC) (2011) Roadmap to a resource efficient Europe. COM(2011) 571 final Brussels. 20.09.2011. Available via DIALOG. https://eur-lex.europa.eu/legal-content/EN/TXT/?uri=CELEX:52011DC0571. Accessed 27 Aug 2019

European Commission (EC) (2013) Green Infrastructure (GI) — enhancing Europe's Natural Capital COM(2013) 249 final. Brussels, 6.5.2013. Available via DIALOG. https://eur-lex.europa.eu/legal-content/EN/TXT/?uri=CELEX:52013DC0249

Fiorini L, Zullo F, Marucci A, Romano B (2019) Land take and landscape loss: effect of uncontrolled urbanization in Southern Italy. J Urban Manag 8:42–56

Grădinaru SR, Hersperger AM (2019) Green infrastructure in strategic spatial plans: evidence from European urban regions. Urban For Urban Green 40:17–28

Gullino P, Battisti L, Larcher F (2018) Linking multifunctionality and sustainability for valuing peri-urban farming: a case study in the Turin metropolitan area (Italy). Sustainability 10:1625. Available via DIALOG. https://www.mdpi.com/2071-1050/10/5/1625. Accessed 27 Aug 2019

Hansen R, Pauleit S (2014) From multifunctionality to multiple ecosystem services? a conceptual framework for multifunctionality in Green Infrastructure planning for urban areas AMBIO 43:516–529. Available via DIALOG. https://link.springer.com/article/10.1007/s13280-014-0510-2. Accessed 27 Aug 2019

ISPRA (2017) Consumo di suolo, dinamiche territoriali e servizi ecosistemici Rapporti 266/2017. Rome, p 196. Available via DIALOG. http://www.isprambiente.gov.it/public_files/RapportoConsumoSuolo2017_rev.pdf. Accessed 27 Aug 2019

Lohrberg F, Licka L, Scazzosi L, Timpe A (2016) Urban agriculture Europe. Jovis Verlag, Berlin, p 230

RAFVGSPB – Regione Autonoma Friuli Venezia Giulia, Servizio Paesaggio e Biodiversità (2018) Relazione Generale al Piano Paesaggistico Regionale del Friuli Venezia Giulia. Available via DIALOG. https://www.regione.fvg.it/rafvg/cms/RAFVG/ambiente-territorio/pianificazione-gestione-territorio/FOGLIA21/#id4. Accessed 27 Aug 2019

Rogers CM, Hiner CC (2016) Siting urban agriculture as a Green Infrastructure strategy for land use planning in Austin, TX. Challenges Sustain 41:39–53. Available via DIALOG. http://www.librelloph.com/challengesinsustainability/article/view/cis-4.1.39/html. Accessed 27 Aug 2019

Rolf W, Peters D, Lenz R, Pauleit S (2018) Farmland – an elephant in the room of urban Green Infrastructure? lessons learned from connectivity analysis in three German cities. Ecol Indic 94:151–163

Sarker AH, Bornman JF, Marinova D (2019) A framework for integrating agriculture in urban sustainability in Australia. Urban Science 3:50. Available via DIALOG. https://www.mdpi.com/2413-8851/3/2/50. Accessed 27 Aug 2019

Secchi B, Viganò P (2011) Une nouvelle question urbaine. In: La ville poreuse. Metispresses, Ginevra, pp 19–22

Termorshuizen JW, Opdam P (2009) Landscape services as a bridge between landscape ecology and sustainable development. Landsc Ecol 24:1037–1052

Viganò P (2012) Idiografia dell'agricoltura. Territorio 60:73–80

Zanfi F (2012) Tra inerzie di figure consolidate e condizioni emergenti. Un ipotesi di lavoro nell'urbanizzazione diffusa dopo la crescita. In: Agnoletto M, Guerzoni M (eds) La Campagna Necessaria, un'agenda di intervento dopo l'esplosione urbana. Quodiblet, Macerata, pp 23–31

Zasada I (2011) Multifunctional peri-urban agriculture – a review of societal demands and the provision of goods and services by farming. Land Use Policy 28:639–648

Chapter 7
The Landscape Planning and the Green Infrastructure in Campania Region

Emanuela Coppola

Abstract In the document *Verso il piano paesaggistico regionale 4.0* (Campania Region, Verso il piano paesaggistico regionale 4.0. Available at http://regione.campania.it/assets/documents/2016-11-versoppr-ver1.pdf, 2016) in the context of the institutional agreement between the Ministry for Cultural Heritage and Activities and Campania Region, it is clarified in the premise that the Cultural Heritage and Landscape Code requires the drafting of the regional landscape plan (PPR), as an organic landscape planning tool extended to the whole Region. The PPR does not give indications of mere protection of landscape assets but, most importantly, defines directives regarding their enhancement, not limited to specific areas and objects. The PPR envisages extending general indications on various and specific assets, from the ones regarding urban areas to the ones regarding agricultural areas and infrastructures, differentiating them for the different territorial areas, based on the overall values of the regional landscape, even overcoming and modifying obsolete constraints. The landscape regional plan (PPR) also has a dynamic role through the possibility to design environmental redevelopment projects aimed at regaining lost landscapes and recreating new landscapes.

The Domitio-Flegreo Coastal Masterplan represents the first regional landscape regeneration project, and green infrastructures become the effective tool for restoring degraded ecosystems.

Keywords Landscape · Green infrastructure · Guidelines · Degraded ecosystem · Landscape regeneration project · Masterplan · Regeneration area · Landscape operational area · Campania Region · Domitio-Flegreo · Coastal Masterplan

---

E. Coppola (✉)
Department of Architecture, Federico II University, Napoli, Italy
e-mail: emanuela.coppola@unina.it

A. Arcidiacono, S. Ronchi (eds.), *Ecosystem Services and Green Infrastructure*, Cities and Nature, https://doi.org/10.1007/978-3-030-54345-7_7

7.1 The Regional Territorial Plan (PTR) and the Guidelines for the Landscape

Despite the recognized importance of the functions of the landscape, in Campania, it is affected by a widespread and increasing degradation, caused by a use of the territory that most of the time does not consider the values that the landscape is capable of expressing in economic, social, cultural and environmental terms. The landscape planning process takes place in Campania through the identification and assessment of landscapes, the definition of landscape quality objectives and safeguard and management policies carried out within an institutional process.

The activities of identifying and evaluating the landscapes, as well as the definition of the objectives and the landscape strategies, take place, on the appropriate scale of analysis, at the different institutional levels. They are developed along a descending flow and an ascending flow. Along the descending flow, the Region defines the reference structural frameworks of the ecological-naturalistic, agroforestry and historical–cultural resources, together with the articulation scheme of the regional landscapes: from these frameworks have been identified the scheme of articulation of regional landscapes.

Along the ascending flow, the municipalities, even in an associated form, the provinces and the metropolitan city can propose, on the basis of the analyses carried out on more detailed scales and the results of the local participation processes, modifications of the reference structural frameworks and of the scheme of articulation of regional landscapes, according to the modalities provided for in article 11 of the Regional Law 16/2004 (flexibility of higher-level planning).

If the landscape plan represents the reference framework for the actions to protect and enhance the landscapes of Campania, the Regional Territorial Plan – in force since 2008 – represents the strategic framework of the territorial transformation policies in Campania. The synergy of the two planning levels must represent the point of reference for any development policy and, therefore, relevant for any community programmes.

The European directives on the environment and the territory are intended as a strong framework for the strategic guidelines of the Regional Territorial Plan of Campania; in particular, the European Spatial Development Planning (ESDP) is considered as a framework for interconnecting territorial diversity. In fact, the PTR interprets the ESPD both as a wide and additional level of supranational planning, as a coordination contribution between community directives and national planning, but also as a connection network between different parts of the European territory.

This method was introduced by the regional law n. 16 of 2004 and implemented by the PTR, which has characterized all the major results achieved in recent years and which leads the municipalities to collaborate more intensively with each other by implementing forms of associated planning (unions of municipalities, associated functions, and administrative federalism) through conferences of areas corresponding to the territorial development systems (STS).

The institutional agreement for the shared drafting of the Campania regional landscape plan was signed on July 14, 2016, by the Minister of Cultural Heritage and Activities and the Campania Region that provided for the following operational steps:

- First macro action – prepare the documents underlying the activity of drafting the regional landscape plan (PPR).
- Second and third macro actions – detection of areas declared to be of significant public interest provided for in articles 136 and 142 of the Cultural Heritage and Landscape Code.
- Fourth macro action – collection of landscape plans existing in Campania.
- Fifth macro action – creation of landscape areas.
- Sixth macro action – preparation of the regional landscape plan (PPR) for the government responsible for the local transformations of the landscape.

The development of the landscape plan is based on the preliminary study of the plan and on the following regional documents: the Guidelines for the landscape and the Charter of the landscapes of Campania, both drawn up in the framework of the Regional Law 13/2008, the Atlas of the landscapes of Campania and the provisions on landscape enhancement contained in the Provincial Territorial Coordination Plan (PTCP).

The preliminary study consists of a report of the following graphs relating to naturalistic values, urbanized areas and urbanization levels, recognition of historical–environmental constraints, from three tables relating of the open rural area (large systems – systems – subsystems) and an analysis of the Vesuvius detailed area.

7.2 Campania Landscapes

With the Campania Landscape Guidelines, the Region applies the principles of the European Landscape Convention to its entire territory and defines the unitary framework for regional landscape planning, implementing the article 144 of the Cultural Heritage and the Landscape Code.

Through the Campania Landscape Guidelines, the Region indicates to the Provinces and Municipalities an institutional and operational path consistent with the principles dictated by the European Landscape Convention, the Code of Cultural and Landscape Heritage and the L.R. 16/04, defining specific directives, guidelines and methodological criteria which compliance is mandatory for the purpose of verifying the consistency of the Provincial Territorial Coordination Plans (PTCP), the Municipal Urban Planning (PUC) and the sector plans, by the respective competent bodies, as well as for the strategic environmental assessment envisaged by Article 47 of the Regional Law 16/04.

The Charter of the landscapes of Campania is composed of the following documents: the Charter of naturalistic and agroforestry resources, the Charter of rural and open land systems, the Charter of historical–archaeological structures and the Scheme of the landscapes of Campania. These documents together form the Charter of landscapes of Campania, constructed and defined as a statute of the regional territory.

They constitute the main reference for the definition of strategies and safeguard guidelines and sustainable management of landscapes and ecological, agri-environmental, historical–archaeological and landscape resources, in accordance with the principles dictated by the Code of cultural heritage and landscape and by the European Landscape Convention.

Table 7.1 Schematic structure of rural and open territory systems that forms the Atlas of Campania landscapes

Large systems	Systems	Subsystems
1. Mountain areas	*Massifs and mountain ranges of the internal Apennine ridge*	1. Matese 2 Taburno 3. Picentini 4. Maddalena 5. Alburni 6. Cervati
	Elevations and mountain complexes of the internal Apennine	7. AltoTammaro 8. Gelbison e Centaurino
	Dorsal and isolated mountain ranges of the preappennine and coastal strip *Internal hill ranges, with clayey lithology*	9. Tifatini 10. Massico 11. Avella, Montevergine 12. Soprano 13. Reliefs of the Sorrento-Amalfi 14. Stella 15. Bulgheria 16. Alto Tammaro and Fortore 17. Alta Irpinia
2. Hilly areas	*Internal hill ranges, with marl and limestone lithology*	18. Medio Volturno 19. Telesina Valley 20 Sabato and Calore 21. Calore Irpino and Ufita 22. Ofanto 23. Avellino Basin 24. Bassa Irpinia 25. Tanagro and Alto Sele 26. Montella
	Hill ranges of the coastal strip, with marly calcareous lithology	27. Salerno and Eboli 28. Calore Lucano 29. Coast of Cilento 30. Internal Cilento
3.Continental volcanic complexes	*Continental volcanic complexes*	31. Roccamonfina Volcano 32. Campi Flegrei 33. Somma-Vesuvius

(continued)

Table 7.1 (continued)

Large systems	Systems	Subsystems
4. Plains areas	*Foothills and terraced plains*	34. Roccamonfina Plain 35. Caserta Plain 36. Phlegraean Plain 37. Vesuvian Plain 38. Nola Plain 39. Irno Valley 40. Sele Plain
	Internal valleys and basins	41. Volturno Valley 42. Monteverna Plain 43. Caudina 44. Diano
	Floodplains	45. Garigliano Plain 46. Volturno Plain 47. Regi Lagni 48. Sebeto 49. Sele Plain
	Coastal plains	50. Garigliano 51. Volturno and Flegrean coast 52. Sarno plain 53. Sele plain
	Volcanic islands	54. Procida Island
5. Gulf islands of Naples	*Limestone islands*	55. Ischia Island 56. Capri Island

The scheme of articulation of the landscapes of Campania represents a first contribution to the identification of regional landscapes (or "landscape areas," in the definition of articles 135 and 143 of the Code of cultural heritage and landscape). The identification of the landscapes is based on the intersection of the studies concerning the material structures of the regional landscape.

The Atlas of Campania landscapes through the Charter of rural and open territory systems identify geographical partitions of the regional territory that are internally characterized by the physiographic aspects of a regional scale that influence sustainable management, productive and ecological potential and the risk of degradation of the resources of the rural and open territory (soils, waters and ecosystems). The Atlas is hierarchically divided into 5 large systems, 12 systems and 56 subsystems, as summarized in Table 7.1.

7.3 The Regional Landscape Areas

In the Campania Region, three types of landscape plans are currently in force:

1. The Landscape Regional Plan (PTP) subjected to the provision of the art. 162 of the Legislative Decree n.490 of 10/29/99 for the following 13 areas: Roccamonfina

Volcanic Group, Matese Mountain Group, Caserta and San Nicola La Strada, Monte Taburno, Camaldoli and Agnano Hill, Posillipo Hill, Phlegraean Fields, Ischia Island, Capri Island, Vesuvius and Monte Somma, Terminio-Cervialto (Picentini Mounts), Cilento Coast and Internal Cilento (Cervati Massif).
2. The landscape plan of the Island of Procida drawn up before the law n. 431 of 1985.
3. The Territorial Urbanistic Plan of the Sorrento-Amalfi area (PUT) approved according to Law 431/85, with the L.R. n. 35/87.

The regional landscapes of Campania are identified not only on the basis of a reading of the material structures (physical, ecological, agroforestry and historical–archaeological structures) but also on in-depth analysis following the leap in scale and above all on the semiological perceptive study that completes the picture of structural interpretation based on the identification of landscapes. Reference documents for the delimitation of the landscape areas were the Landscape Charter, the Landscape Atlas, the PTR and the Guidelines for the Landscape.

Not all the elements and constitutive relationships of material structures have the same weight in identifying landscapes: the need to prefigure a series of landscape areas with their own identity, therefore, a defined spatial structure, even if with large overlaps, involves a greater attention to the convergence of those ecological and historical–archaeological systems considered significant with respect to a geomorphological structure recognizable as unitary both from inside and from outside.

In fact, the landscape areas intersect with the areas of the open rural territory and with the territorial development systems (Table 7.2).

7.4 Landscape Enhancement Through Provincial and Municipal Plans

Regional planning has as its main objective the government of territorial transformations according to the principles of clarity, certainty of administrative law and effectiveness of administrative action in a framework of efficiency in the use of resources. In accordance with the principle of subsidiarity, the planning system is divided into three levels: regional, provincial and municipal.

The three types of plans regulate the transformations of the territory through different processes: the PTR dictates the addresses of the entire regional territory, the Provincial Territorial Coordination Plans (PTCP) determine the structure of provincial territory and the PUCs deal with the management of municipal area. The outlined planning framework thus defines the physical aspect of planning.

At the same time, there is a more complex picture of the plans of landscape and territorial value in force, which interact and overlap with them finding the synthesis within the regional landscape plan (PPR).

For these reasons, the Campania Region will implement its landscape planning also through a system of parks, nature reserves and the Regional Ecological Network

Table 7.2 Landscape areas and territorial development systems (STS)

Landscape areas	Material structures of the historical–archaeological landscape	Territorial development systems (STS)
Alto/Basso Garigliano	*System of pre-Roman fortified centres/ Agro centuriato of Minturno*	A11 F1 (C6)
Domitian coast	*Archaeological sites of Liternum*	F1 (E4, C8, F2)
Carinola and Volturno Plain	*Ager Falernus*	C6 (F1, D4, E4, A11, B7)
Matese	*System of pre-Roman fortified centres*	A10 (B6)
Phlegraean Plain/Campi Flegrei	*Centuriation of Capua/Campi Flegrei archaeologic system*	C8, E2, E4, F2 (E1, D3, C8)
Islands of Ischia and Procida	*Archaeological site of Vivara*	F5, F2
Napoli	*Historical centre of Naples*	D3 (E2, C7, F3, F2, C8)
Casertano	*Archaeological and agro-centuriate system of Capua*	D4, C6, B7 (E1, A9, E4)
Acerrano	*System of Greek–Italic archaeological sites*	E1 (E3, D4)
Vesuvius	*Pompeii archaeological and agricultural sites*	C7, F3, E3 (E1, D3, C5)
Beneventano	*Benevento Centuriation*	C2, B5, B4, B3, D1 (B6, A9, A12)
Nolano	*Agro centuriato and historic centre of Nola*	E3, B8
Avellino Basin	*Avellino Centuriation*	D2, C3 (A12, A8)
Capri Island	*Historical and cultural landscape of Capri*	F4
Amalfi Coast and the Lattari Mountains	*Cultural landscape of the Amalfi Coast*	F4, F7 (C5, D5, F3)
Salerno	*Historic centre of Salerno*	D5 (C4)
Picentini Mountains	*System of Roman archaeological sites*	A7 (D5, B2, C4)
Alto Tanagro/Tanagro Valley	*Archaeological site and Centuriation of Volcei*	B2, A1 (F8, B1)
Sele Plain/Diano Valley	*Cultural landscape of Cilento*	A1, A2, A3, A4, A5, B1, F8, F6, D5, A7 (A6, B2)
Partenio Sanctuary of M. Vergine	*Neolithic archaeological sites*	B8, E3, A8 (D4, A9, D1, D2, E1)

(RER), starting from the consideration that natural landscapes and human landscapes are closely interrelated, so interventions aimed at the maintenance or upgrading of natural environment take on the role of reconstruction and redevelopment of anthropic landscapes and active conservation of landscapes in general. The PTCPs currently in force are the PTCP of Salerno, the PTCP of Caserta and the PTCP of Benevento in force since 2012; the PTCP of Avellino, in force since 2014; and the PTCP of Naples still in progress. The following picture is a summary of the PTCP landscape components (Fig. 7.1).

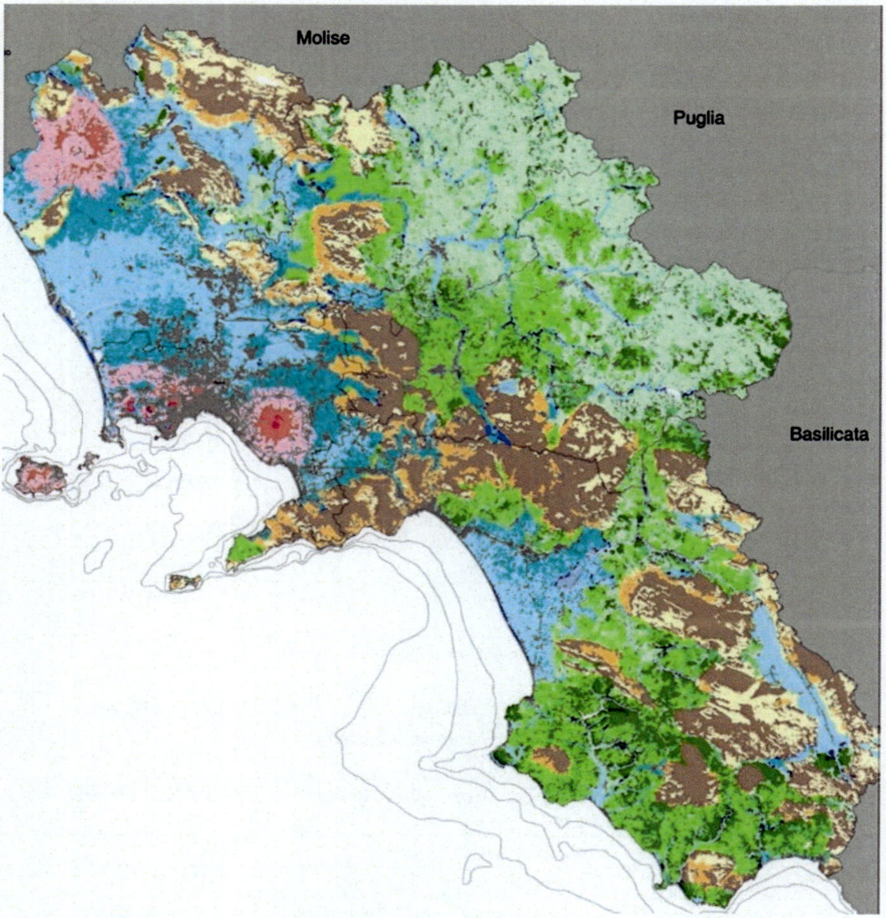

Fig. 7.1 Scheme of the landscape components of the PTCPs. (Source: Campania Region 2016)

The landscape plan from a procedural point of view is structured according to phases. The first phase is the drafting of the preliminary plan, useful for starting strategic environmental assessment and public consultations. The preliminary will be elaborated for the whole Region identifying the operational areas. These areas will be shared by the municipalities in the consultation phase foreseen for the preliminary. The operational areas, which together constitute the totality of the regional territory, are the fundamental instruments for the preparation and approval of the PPR (Fig. 7.2).

They are bounded on the basis of administrative boundaries taking into account the areas of landscape and areas of the parks. "Cartographic windows" have been introduced when an operational area must necessarily cut mountain or volcanic

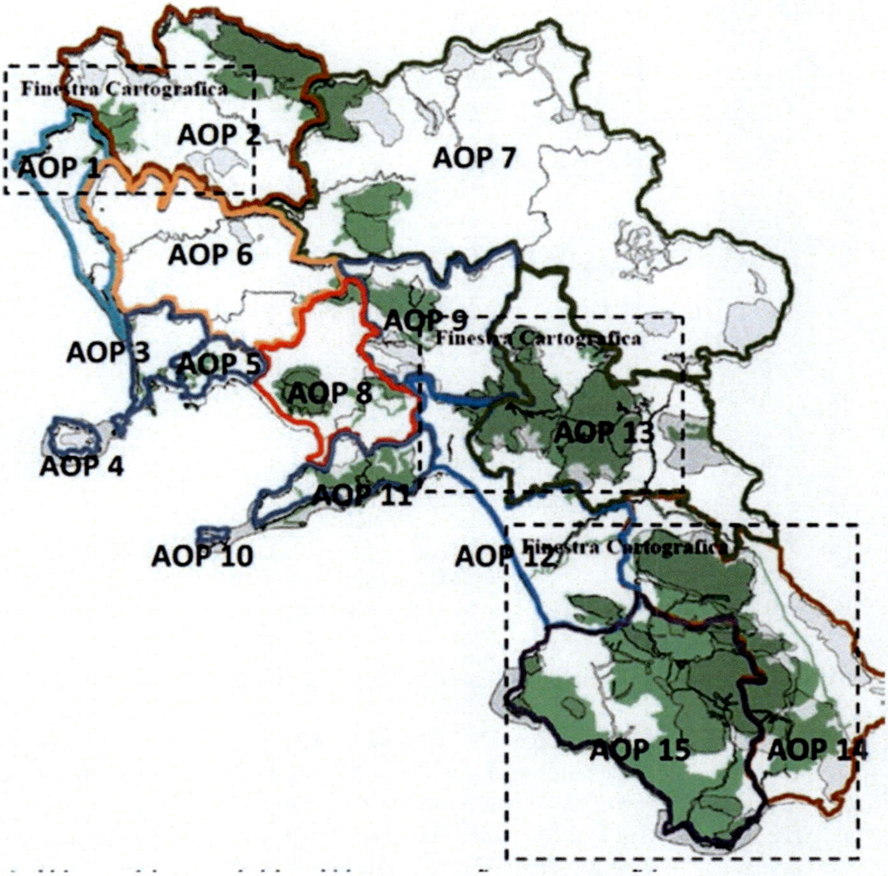

Fig. 7.2 Landscape operational areas, park areas and cartographic windows. (Source: Campania Region 2019)

areas (generally park areas). The operational areas include multiple landscape areas that are complementary to each other.

A first hypothesis of these areas is shown in Table 7.3.

7.5 The Regional Landscape Regeneration Projects

The PPR also has a dynamic role through the possibility to design environmental redevelopment projects aimed at regaining lost landscapes and recreating new landscapes. Projects that could be of environmental restoration (Coppola 2016; Moccia 2013; Moccia and Coppola 2013) where the historical–geographical reading of the

Table 7.3 Operational area (AOP) and landscape areas

n. AOP	n. Landscape areas	Landscape areas
1	2 – 3 – 4 – 5	Garigliano, Domitian Coast, Volturno Plain
2	1 – 6 – 7 – 8	Alto Garigliano, Roccamonfina Volcano, Medio Volturno, Matese
3	10 – 11 – 12	Flegrea plain, Campi Flegrei, Procida Island
4	12	Ischia Island
5	13	Napoli
6	14 – 15	Casertano, Acerrano, Titerno
7	9 – 17 – 18 – 19 – 20 – 21 – 32	Taburno e telesina Valley, Fortore e Tammaro, Beneventano, Ufita, Caudina Valley, Baronia
8	16 – 22 – 24	Vesuvius, Nolano, Sarno Plain
9	23 – 49	Avellino, Partenio
10	25	Capri Island
11	26	Amalfi Coast and the Lattari Mountains
12	27 – 28 – 37	Irno Valley, Salerno, Sele
13	29 – 30 – 31 – 33 – 34 – 51	Picentini, Terminio
14	35 – 36 – 38 – 39– 40 – 43 – 47 – 48 – 50	High Tanagro, Tanagro Valley, Calore, Diano Valley, Gelbison Cervati, Bussento, Policastro
15	41 – 42 – 44 – 45 – 46	Stella, Alento, Pisciottano, Bulgheria, Mingardo

territory represent "the analytical phase necessary to implement the principles of planning with nature."

These regional landscape regeneration projects are provided for areas of particular importance due to the landscape fragility and the degradation conditions (strong diffusion of illegal phenomena) and represent the weak and compromised part that the PTR, with the Regional Law 13/2008, classified as Complex Territorial Fields (Ctc) or areas with a greater degree of vulnerability.

These also represent identity landscapes as well as places of particular criticality, and they are as follows:

- Ctc 1–2–3–11–12 (Basso and Alto Casertano – Regi lagni – urban area of Caserta – north Naples/Caserta – acerrano-giuglianese area)
- Ctc 4–5 (Alto sannio-irpino – interprovincial area Caserta/Benevento/Avellino – avellinese area)
- Ctc 6 (Salerno coast – Internal and coastal Cilento)
- Ctc 7 (Sorrento-Amalfi peninsula – Capri Island)
- Ctc 8 (Domitian coast)
- Ctc 9 (Vesuvian area – Sarno plain – Nola Territory)
- Ctc 10 (Campi Flegrei – Ischia and Procida Islands)

The projects are aimed at the regeneration of the urban fabric, the reorganization of the settlement plot, the territorial redevelopment and the enhancement of the landscape. Moreover, they point to restore environmental, urban and architectural features of the built-up area; to increase the attractiveness and competitiveness of the territories; to improve the qualitative and quantitative level of equipped public spaces; to raise the architectural quality of the built environment, also through the use of bio-architecture and eco-compatible materials; and to propose methodological and innovative approaches to support planning choices, aimed at contributing to a more efficient and transparent territorial governance.

7.6 The Domitio-Flegreo Coastal Masterplan and the Green Infrastructure Project

The Masterplan, developed by Andreas Kipar in 2018 as the winner of a competition of the Campania Region, represents the first regional landscape regeneration project (Fig. 7.3). The networks, and first of all the regional ecological network (RER), constitute the reference point for the integration of local and sectoral policies in the broader context of regional policies.

In particular, through the construction of the ecological network at different levels (regional, provincial and local), the concrete possibility of developing active policies for the protection of environment and landscape is manifested, involving also sectorial planning. The ecological network is configured as a programmatic tool that allows to achieve an integrated management of resources and the regional physical–territorial space, including the landscape.

Furthermore, by extending the objectives of improving environmental quality, conservation and increasing biodiversity to the entire regional territory and linking them to landscape quality objectives, the RER becomes the instrument through which sustainable development policies make concrete use of the contribution of more subjects that cooperate to achieve those objectives, starting from the local and sectorial realities.

In this context, the Litorale Domitio Flegreo itself becomes a model. Despite the obvious deterioration, the development potential of this area is rather remarkable.

> The coast, for the most part covered by the beautiful pinewood pine, is still an extraordinary resource. Rivers, canals, wetlands and lakes characterize this water landscape. There are also important cultural, historical and archaeological emergencies. A veritable widespread open-air museum. An exceptional context that requires the setting of complex processes based on a radical paradigm shift. The landscape, from an object of exclusive protection – which has not guaranteed its conservation, becomes an engine of sustainable development. (Kipar 2019)

The green infrastructures, promoted by the European Union since 2013, become the effective tool for restoring degraded ecosystems, also taking advantage of the opportunities offered by European funds (Coppola and Vanella 2016; Coppola et al. 2019).

The objective is the activation of new productive landscapes through a structured programme of public and private interventions coordinated by an overall project vision, of which the Masterplan is the implementation tool. Projects that express environmental and landscape quality as the source of development and regeneration

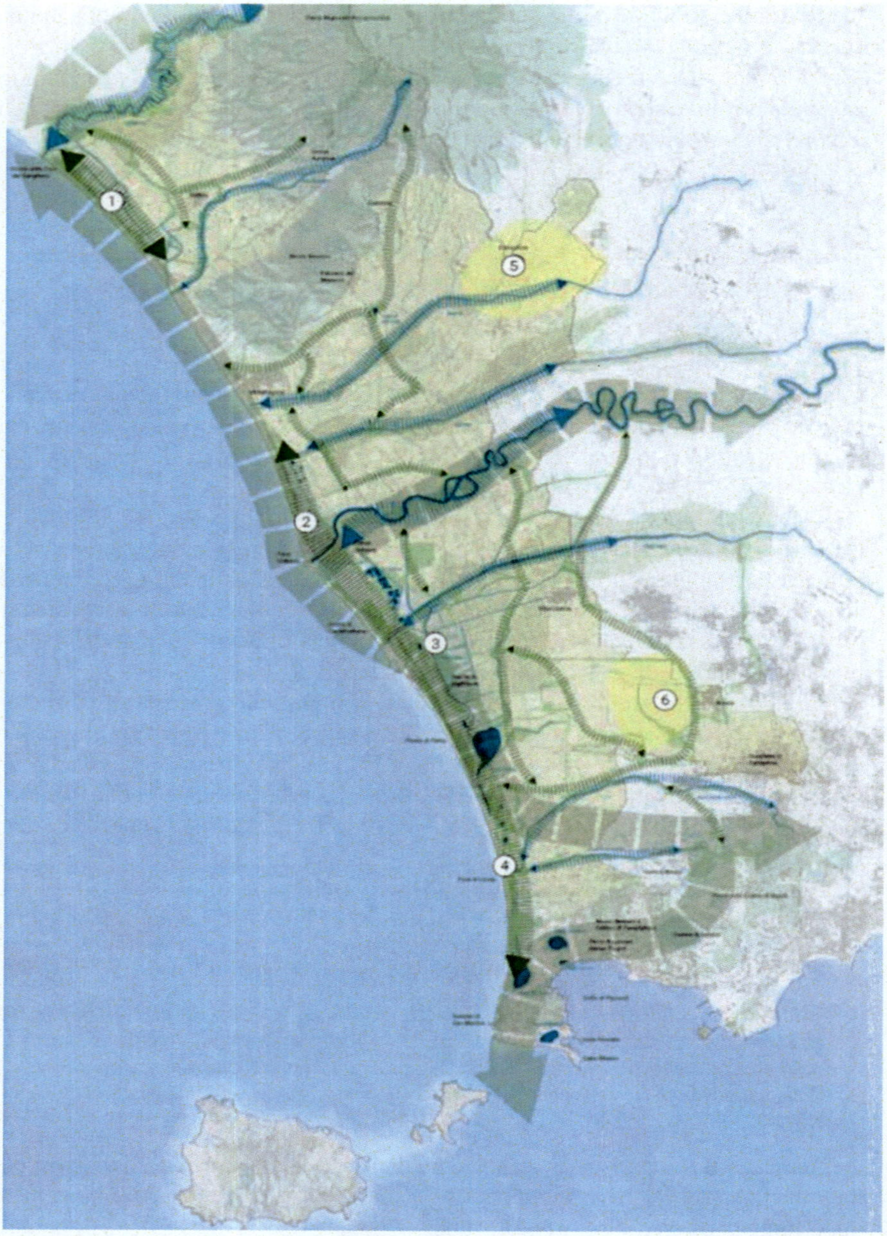

Fig. 7.3 Green infrastructure in Domitio-Flegreo Coastal Masterplan. (Source: Campania Region 2019)

and that will allow us to rediscover new productivity in line with the vocations and values that the territory expresses.

A future that also sees the landscape as an opportunity for tourism: a quality tourism interested, in addition to the pleasures of bathing, also to the enjoyment of natural and cultural beauties, and to the relationships with the local agricultural system that finds a hospitable place that can accommodate and make the many exceptional goods easily accessible and usable.

The Masterplan proposes a vision based on three main strategies:

1. The requalification and enhancement of the ecological and landscape–environmental system, in order to create an articulated system of territorial permeability and ecological connectivity.
2. The enhancement of the historical–cultural and agricultural system, through the improvement of the fruition offer and the creation of quality places where to expand the offer of agro-environmental products and services and eno-gastronomic tourism.
3. The development of a sustainable and integrated mobility through the enhancement of rail transport offer, supported by an improvement of transversal connections and by the new role of stations as intermodal polarities and the provision of an articulated system of cycle routes set on a backbone that connects the whole coast, becoming a completely innovative mobility offer for the territory.

Three strategies to which are added the many public and private projects currently being evaluated that will constitute the programme of interventions compatible with the active protection of the territory and that must make synergy between them.

7.7 Conclusions

Green infrastructures become an effective tool for the regeneration of large territories affected by profound imbalances, compromised by decades of inattentive development like that of the Domitio-Flegreo coast, which at the same time has potential linked to the presence of water, nature and agriculture, as well as a very rich historical and cultural heritage.

The theme is declined in two different strategic paths: the system of linear green infrastructures that follow the natural routes of watercourses and natural parks and a series of specific areas where agricultural and environmental innovation could be experimented.

A complex and articulated process that for this reason will need initial actions of great symbolic value, around which it will be possible to generate attention and interest to participate in citizenship.

References

Campania Region (2016) Verso il piano paesaggistico regionale 4.0. Available http://regione.campania.it/assets/documents/2016-11-versoppr-ver1.pdf

Campania Region (2019) Litorale Domitio-Flegreo. Strategie per la rigenerazione territoriale, ambientale e sociale. Available: http://porfesr.regione.campania.it/it/progetti-e-beneficiari/masterplan-litorale-domitio-flegreo

Coppola E (2016) Infrastrutture sostenibili urbane. INU Edizioni – Collana Accademia, Roma, p 213

Coppola E, Vanella V (2016) Infrastrutture verdi e morfologie urbane. Una proposta metodologica per l'area metropolitana di Napoli. Urbanistica 157. INU Edizioni– ISNN 0042-1022

Coppola E, Cirillo C, Youssef R et al (2019) Ameliorating a complex urban ecosystem through instrumental use of softscape buffers: proposal for a green infrastructure network in the metropolitan area of Naples. Front Plant Sci Sect Crop Prod Physiol. https://doi.org/10.3389/FPLS.2019.00410. Available: http://journal.frontiersin.org/article/10.3389/fpls.2019.00410/full?&utm_source=Email_to_authors_&utm_medium=Email&utm_content=T1_11.5e1_author&utm_campaign=Email_publication&field=&journalName=Frontiers_in_Plant_Science&id=439215

Kipar A (2019) Masterplan Litorale Domitio-Flegreo. Strategie per la rigenerazione territoriale, ambientale e sociale. Available http://regione.campania.it/assets/documents/2016-11-versoppr-ver1.pdf

Moccia FD (2013) ecological restoration methodology. In: Moccia FD, Palestino MF (eds) Planning stormwater resilient urban open spaces. Clean, Napoli, pp 12–17

Moccia F, Coppola E (2013) Si può parlare di pianificazione paesaggistica nelle aree ad elevata antropizzazione? In: Petroncelli E (ed) Il paesaggio tra rischio e riqualificazione. Liguori Edizioni, Napoli

Chapter 8
The Green Infrastructure Instrument for the Metropolitan Area of Naples: Experimentations Through Local Planning

Francesco Domenico Moccia and Antonia Arena

Abstract Climate changes produce harmful effects on physical, biological and human system, particularly on human health; the possibility to reduce negative impact of phenomena as the urban heat island, the frequency, intensity and duration of extreme precipitations, the heat waves and cool waves, depends on the capacity to plan urban settlements improving their resilience. The green infrastructure planning and the assessment of urban quality in relation to ecosystem services are affirming as outcomes of studies and researches in urban planning. The Municipal Master Plan of Qualiano is an experimentation of green infrastructure planning according to an ecosystem approach aimed to improve ecosystem services supply at a local scale. Furthermore, it is a significant case of Metropolitan Ecological Network realized by a bottom-up approach: despite the lack of a metropolitan plan, in fact, it is possible to evaluate ecosystem values of landscape, both natural and urban, and to plan them with an integrated approach to increase ecosystem services.

Keywords Climate change · Ecosystem services · Green infrastructures · Metropolitan planning · Multi-scalar approach · New urban standards · Regeneration processes · Ecological restoration

8.1 Green Infrastructures and Ecosystem Services

8.1.1 New Issues for Urban Planning

New important issues are highlighted in the planning field under a scientific point of view; they are born from economic and social, ecological, environmental, technological and energetic changes. In urban planning field, significant issues are topics

F. D. Moccia (✉) · A. Arena
Department of Architecture, Università degli Studi di Napoli Federico II, Naples, Italy
e-mail: fdmoccia@unina.it

A. Arcidiacono, S. Ronchi (eds.), *Ecosystem Services and Green Infrastructure*, Cities and Nature, https://doi.org/10.1007/978-3-030-54345-7_8

related to the urban settlements' quality and security and the territory capacity adaptation to the climate change effects (Moccia 2009). Two important issues are in the mainstream: on one hand, the need to maintain or restore the balanced link between both environmental and urban systems, on the other hand, the requirement of a qualitative approach able to evaluate efficiency and performance of urban spaces (Giaimo and Barbieri 2018; Moccia 2014). In this frame, green infrastructures are becoming a paradigm to define not only the structure of environmental networks, but also the urban settlements analysed and interpreted through relationship of both public and private spaces (Angrilli 2016a; Arcidiacono et al. 2018).

In Italy, since the 1970s, the public spaces provision was ensured by law about standards that rule the quantity of areas for public facilities, education, parking and green public spaces in area quantitative extension; nowadays, we need an integrated approach that takes quality into account, assessing ecosystem services supplied within urban environments (Barbieri 2008).

Ecosystem services are assessed with a "customer-oriented" approach, that is, from a point of view oriented to the beneficiaries (Cortinovis and Geneletti 2019); therefore, they can help to have a new approach to the aforementioned standards of the Italian planning system.

Green infrastructures provide ecosystem services. They suggest a new paradigm in territorial and urban planning because they have multifunctional features, that is, spaces with both environmental and urban functions: on the one side, green infrastructures link natural areas, increase biodiversity, improve air quality, reduce noise and atmospheric pollution and soil waterproofing and restore natural cycles as water cycle or biotic cycle; on the other, in relation to urban functions, they ensure the presence of natural rural areas, public green and social spaces, promote sustainable mobility and energetic and digital networks (Arcidiacono et al. 2018). Green infrastructures become a natural and semi-natural structure of anthropic settlements. Furthermore, they are able to synchronize natural elements with anthropic ones to improve environment quality (Angrilli 2015, 2016a, b).

"Green Infrastructure can be broadly defined as a strategically planned network of high quality natural and semi-natural areas with other environmental features, which is designed and managed to deliver a wide range of ecosystem services and protect biodiversity in both rural and urban settings" (European Commission 2013:7). They can be drawn and implemented along landscape linear elements; they can replay natural process through new technologies and through nature-based solutions to provide public services and common benefits (Arena 2017; Coppola 2016).

About the issue of settlements, assessment in function of performance and quality, in Italy, planning standards answered to complex social claims; in addition, their quantification by law helped to carry out public services with efficient results (Salzano 2007). An evaluation about post-war urbanism in Italy led to recognize that a quantitative approach has prevailed in city planning practice.

In urban and territorial planning, a switch from quantitative to qualitative approach is needed to focus on provided services rather than the simple quantity of public spaces. Increased consideration of qualitative aspects allows to pursue an integrated approach, which considers both natural and environmental features and socio-economic ones, holding together the different sustainability components (Garramone and Gissi 2018); public spaces become an efficient standard to assess the urban quality (Oppio et al. 2018).

8.1.2 Ecological Networks Among Regional, Metropolitan and Local Planning

The establishment of Metropolitan City by law n. 56/2014 has rekindled the scientific debate on their functions and on the role it can play (Barbieri 2015; d'Alessandro and Realfonzo 2018; De Luca and Moccia 2017): among them, we can stress their power, by the new law, in strategic and territorial planning.

In Campania, the constitution of Metropolitan City, a new local government and a new planning level, interrupted over and over again[1], of the Metropolitan Plan, characterized by innovative elements in relation to the time it was proposed, could be review and update, introducing new topic, issues, approaches and solutions coming from technical and scientific progresses.

Metropolitan dimension is the most adequate to plan ecologic connections and to upgrade landscape because of the link between regional and municipal planning, leading to local administration activities. The Metropolitan Plan defines general guidelines, through an integrated analysis of territories, understands and plans landscape elements which could appear fragmented and without potentiality at a local scale. In Campania, the Regional Territorial Plan underlines that in Metropolitan City of Naples, high-density settlements are interrupted only by residual green space. Flegrea area, Vesuvio and monte Somma orographic system, Sorrento peninsula's coast and Gulf islands are the most relevant and better preserved environmental, cultural and landscape sites; the Metropolitan Ecological Network (MEN) (Fig. 8.1) can be drawn by linking these areas.

Also, if the PTCP, the Territorial Cohordination Provincial Plan, is not approved by the Metropolitan Council, but only by the Giunta, its strategic lines are guidelines for municipal urbanism and are improved in the inter-institutional planning cooperation.

The MEN was one of the most innovative planning decisions proposed by Territorial Coordination Plan of Naples Province, never approved. According to the regional approach, the MEN should aim to achieve two main goals: on one hand to protect, conserve and connect high-level quality biodiversity areas, on the other hand, to provide open areas for cycling, hiking and other amenities.

Despite the lack of metropolitan guidelines, the construction of green infrastructures is an important objective to improve the quality of life in urban settlements. Therefore, strategies and actions plans at local levels become more relevant (Moccia and Berruti 2018). At the metropolitan scale, green infrastructures are a simple

[1] To clarify further cf. https://www.cittametropolitana.na.it/pianificazione_territoriale/ptcp

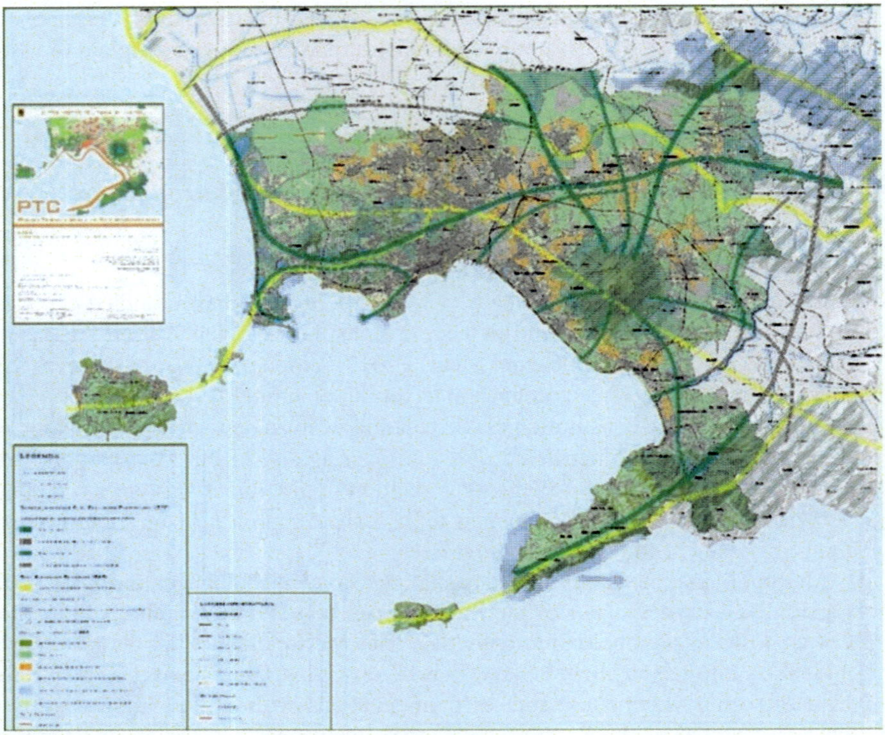

Fig. 8.1 Ecological network at metropolitan level (Source: PTC 2011)

sketch of connecting directions without a clear reference to geographic sites selected on the basis of their ecologic qualities. In other words, these lines drawn on the map of PTCP express objectives of linking nodes of high natural and biodiversity high value. They do not offer more information on the way linkages may be made. In this way, at the municipal level, there is an opportunity to plan in more detail on the basis of land biodiversity, functional and biological state of creeks and rivers and ecological fragmentation.

A further step may be made by metropolitan or inter-municipal plans, in case their extension is that of a hydrologic basin where a whole water cycle can be managed, being one of the most important factors of life in a geographic area (Moccia and Scalera 2018). But in this chapter, we concentrate on the following steps because it seems the most fertile we find in our planning practice.

In fact, each municipality can help to construct metropolitan green infrastructures, promoting the restoration of the space, ecological quality and ecological links; also, municipalities characterized by urban character – business, manufacturing and directional – must give attention to environmental aspects and to improve landscape quality, within own plans (Moccia 2013a, b).

8.2 Green Infrastructures and Ecosystem Services in Qualiano's Municipality Plan

8.2.1 Territorial Contest

The Municipal Master Plan of Qualiano had the scientific support of a working group[2] of the Department of Architecture of University of Naples Federico II in a process involving the political and technical bodies of the municipality.

Qualiano Municipality is located on northwestern border of the metropolitan area of Naples, developed since the second post-war period, mainly for the migrant flows coming from city centre, in a typical suburbanization process. About 25,000 people live in Qualiano on a surface of 7.26 sq. km. The population is younger than metropolitan average and the major part of workers are occupied in building and trade sectors.

Qualiano is placed in the Camaldoli hydrographic basin whose watercourse runs on the municipal territory, close to city centre and alongside the edge of Ripuaria road. Local hydrographic network is composed of other watercourses, as Cavone Croccone, which runs parallel to Camaldoli river, and many other short waterways, some of whom are also artificial.

City centre, localized in the southern part of the municipal territory, is welded to Villaricca urban area, one more metropolitan suburb; while, at the northern side Circumvallation of Naples flows, the expressway built in the second half of the 1900s, along which productive and commercial activities are developed, making a long strip join most of the residential suburbs of the north-western metropolitan sector. The remaining portion of territory has an agricultural character, although it lacks economic relevance. Nevertheless, on the land, we can still find the sign of agricultural divisions and rivers tracks.

Historical streets and orthogonal agricultural pattern constitute the orientation elements and have guided the growth of the urban centre. This has grown through parcelling: within regular grids of roads and lots mostly saturated with housing while public equipment and public spaces are lacking, according to the most widespread practice of contemporary urban planning in southern Italy. Nowadays, among the four types of planning standard required by law (public facilities, educations, parking and green spaces), only areas used for schools fulfil minimum standard thresholds; on the contrary, public and leisure open spaces are lacking, except one public park, recently enlarged, which has become suddenly very popular and which is, in any case, insufficient to supply the needs of the whole population; finally, parking areas and cultural and social facilities are lacking, as shown in Table 8.1.

Another special element is Ripuaria street, traced alongside Camaldoli river, around which urban sprawl settlements were developed in the absence of urban plan

[2] A. Arena, A. Sgobbo, A. Nigro with the scientific responsibility of F.D. Moccia.

Table 8.1 Current and planned state of standard areas

	Public facilities (sq. m.)	Educations (sq. m.)	Parking (sq. m.)	Green spaces (sq. m.)
Current	29,236	34,210	5044	41,338
Deficit[a]	22,130	81,364	59,164	189,809
Planning	**22,564**	**82,583**	**59,340**	**246,411**[b]

[a]Population at 31.12.2017 (Source: ISTAT)
[b]Data include areas designated to park river and inter-municipal park

or derogating from the actual ones (Arena 2018). The increase of impervious soil after housing and infrastructure construction heightens the hydrological risk related to more quantity and higher speed run-off and the presence of exposed values as goods and people. The lack of facilities and public space, the low quality of environment and the management of environmental risk are the main critical issues of Qualiano's territory.

According to the Campania Region urban planning rules, Municipal Urbanism Plan (MUP) is divided into two documents: structural and operational. In the first, there are rules to preserve natural and cultural goods, while in the second, there are projects of city change or regeneration. In such documents, green infrastructures and public spaces are indicated as goods to preserve (structural plan), but will be also objective of improvement, naturalization and fruition in the operational part finalized to provide ecosystem services according to a sustainable approach; in this way, these solutions to improve urban resilience are introduced into territorial and urban planning.

8.2.2 The Municipal Urbanism Plan and the Environment Regeneration Goals

The MUP aims, as main goals, to put stable urban settlements in order, to equip the city with public spaces and facilities and to improve urban and environmental quality. MUP and the possible development projects are based on three foci (Fig. 8.2): (1) public spaces are the structural elements of urban settlements; (2) natural and environmental resources will be interested from environmental requalification, restoration, renovation and fruition and they will be integrated into urban settlements to provide ecosystem and urban services; (3) urbanism and infrastructural facilities are the flywheel to develop existing economic activities and to attract new ones.

The first goal is pursued in the residential areas where public spaces are needed, through urban regeneration projects, while in the new development zones, they are already decided in the MUP and indicated for future development, so that there will be a clear connection with the urban public space of the whole suburb. The second goal is based on the river landscapes restoration (Arena 2017; Moccia

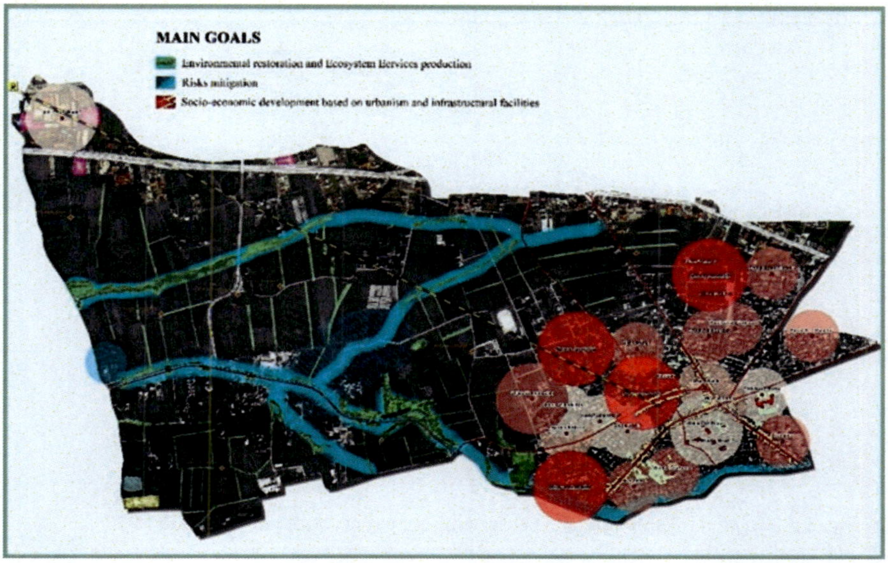

Fig. 8.2 Main goals and focus into Qualiano MUP

2014) of biological natural cycles and services and integrating them with urban uses. Finally, the third focus point is implemented in developing productive and commercial activities along SP1 expressway to strengthen already existing functions through the rule that at least 30% of new buildings must be for non-residential functions.

In Qualiano MUP, public spaces design becomes a relevant element to ensure minimum thresholds of public services, to improve urban quality and to guarantee the right to the city (Lefebvre 1968; Indovina 2018). Green infrastructures design of the plan has three different levels – metropolitan, peri-urban and urban – with different characters and goals. At the metropolitan level, a river park is proposed along two riverbanks that run in parallel and are connected by a park strip; the planned natural network touches the urban area through peri-urban landscapes where MUP permits agritourism destinations to make the agricultural sector most profitable and to strengthen the buffer zones among the most natural areas and urban ones; finally, within urban areas, a corridor of open green spaces – with different functions – is devised to ensure the continuity of green infrastructures. Moreover, the plan promotes the resilient regeneration of public and private heritage (Fig. 8.3).

Other MUP regulations have the objective of improving settlement sustainability as higher density or volume building incentive for near zero emissions constructions and sustainable urban drainage solutions (Berruti et al. 2013).

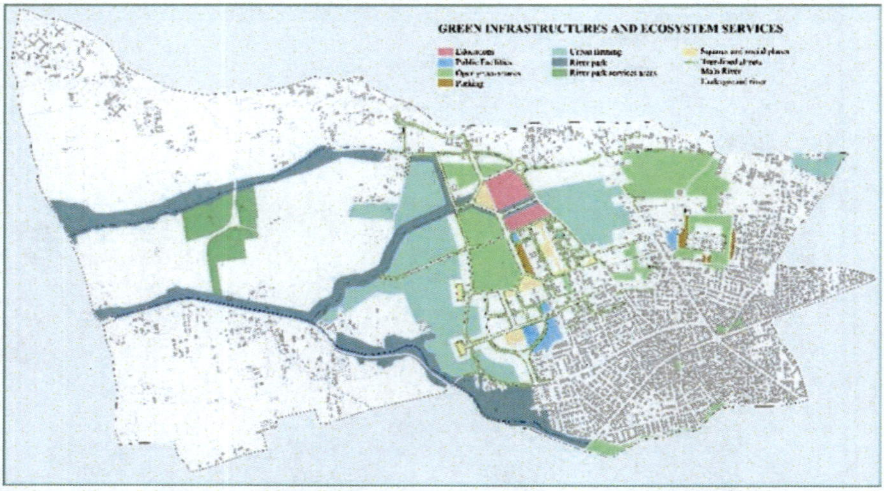

Fig. 8.3 Green infrastructures and ecosystem services network into Qualiano MUP

8.2.3 Details of Green Infrastructures and Ecosystem Services Design

The river park boundary is designed on the existing tracks of natural elements linked to the river (for example river vegetation). A fixed buffer would not have taken into account the current site state and their possible development; therefore, the park has a flexible section, as its functions and uses are variable. In this way, the river park has double relevance: (a) metropolitan role, because it improves links among high-level natural areas providing first-level ecosystem services; (b) inter-municipal one, since it provides cultural and leisure services for inhabitants and resident who come from close municipality. River park increases, on one hand, provisioning services such as soil, water and biodiversity, and, consequently, regulates the microclimate and the water management, and, on the other hand, gives recreation services with greater benefits of physical and mental health of people. Furthermore, green infrastructures design deals with the management of risk issue (Novotny 2007; Moccia and Sgobbo 2016) within vulnerable areas developed through spontaneous processes (Arena 2018) in which streets or urban settlements have modified natural riverbeds.

Peri-urban areas, as mentioned above, play a role of buffer zone between rural and urban areas and are destined, by plan, for agriculture conservation: the plan permits uses such as urban vegetable gardens, holiday farms – in which production is integrated with trade – educational farms, accessible and usable from different kinds of users. In fact, these kinds of uses help in the management of open, natural spaces, encouraging farmers to stay over land.

Streets are not only a circulation device, but main places in which collective life takes places; therefore, they have a multifunctional character: MUP, inspired by nineteenth century boulevards, designs wide tree-lined public spaces, with cycling–pedestrian lane and water-sustainable management systems. According to an eco-systemic point of view, trees improve shading, capturing particulates and reduce urban heat island effects. The 'sports citadel' and the 'school citadel' – public urban equipment – are located alongside such boulevard-type network: the former in expansion of existing sport facilities, the latter made of a school complex with primary and secondary schools.

Also when public facilities are decided in a MUP and related to a single municipality, it should be taken into account that in metropolitan areas there is a high level of interchange among close municipalities, and users of the before-mentioned facilities will be calculated not only in the MUP municipality. Therefore, planning goals, which improve urban conditions and strengthen ecosystem services by increasing biodiversity, restoring autochthon species, reducing soil waterproofing, introducing sustainable water and waste management, reducing pollution and improving urban microclimate, have always a metropolitan dimension.

8.3 Conclusion

The experimentation carried out in the Qualiano MUP planning process shows the multilevel work to develop ecological metropolitan network through the implementation of green infrastructures and ecosystem services.

Development choices of urban level can help to conserve natural reserves and to support ecological metropolitan network: by designing local streams or rivers as green infrastructures, there is a contribution to join larger ecological corridors through a constituting tract of them. In this case, the Qualiano MUP has identified a tract of the links among regional parks of Campi Flegrei, Metropolitan Park of Naples Hills and the land part of marine reserve of Licola. At the municipal level, what is a simple direction indication of ecological corridors becomes a physical complex of geographical and biological components, able to evolve in more equilibrated ecosystems, really constructing the biological linkages indicated as objective in the upper level plans. Furthermore, green infrastructures in urban areas improve the quantity of public facilities and benefit people living in the suburbs supplied by sustainable open spaces.

Planning networks of green infrastructure is one of the most important parts of the MUP; they are made of different spars and nodes, each one with specific goals, but all together collaborating to the scope of providing ecosystem services and ensuring linking among high-level biodiversity areas. River parks and tree-lined streets, showed in this essay, are the link between both metropolitan parks and urban green spaces. The presented approach introduces relevant changes in the approach to urbanism, already focused over zoning and regulation, because it stresses the issue of infrastructure as a main tool to operate urban regeneration.

References

Angrilli M (2015) Infrastrutture verdi e blu. Urbanistica Informazioni 263 s.i.: sez V, 1–2

Angrilli M (2016a) Definizioni e ruoli delle infrastrutture verdi e blu. In: Moccia FD, Sepe M (eds) Reti e infrastrutture dei territori contemporanei. INU Edizioni, Roma, pp 176–186

Angrilli M (2016b) Spazio pubblico e prestazioni ecologiche. In: Berruti G (ed) Esplorazioni urbanistiche dello spazio pubblico. INU Edizioni, Roma, pp 128–139

Arcidiacono A, Giaimo C, Talia M (2018) Priorità per il progetto urbanistico. Reti ambientali, nuovi standard e rigenerazione urbana. In: Moccia FD, Sepe M (eds) Sviluppare, rigenerare, ricostruire città. Questioni e sfide contemporanee. INU Edizioni, Roma, pp 166–179

Arena A (2017) Pianificare paesaggi lineari. Il restauro paesaggistico delle infrastrutture ambientali e antropiche nei piani territoriali ed urbanistici. PhD Thesis. Available at: http://www.fedoa.unina.it/11563/

Arena A (2018) Spontaneous urban areas planning: experimentation between plan and urban design. TRIA 2:101–118. https://doi.org/10.6092/2281-4574/6086

Barbieri CA (2008) Costruire la città pubblica: nuovi bisogni e nuove risposte. In: Galluzzi P, Vitillo PG (eds) Rigenerare le città. La perequazione urbanistica come progetto. Maggioli Editore, Santarcangelo di Romagna, pp 91–94

Barbieri CA (2015) Dall'istituzione all'azione della Città metropolitana di Torino: il ruolo di una nuova pianificazione. Il Piemonte delle Autonomie:8–14

Berruti G, Coppola E, Moccia FD (2013) Urban morphology and ecological water-centered design in Mediterranean areas. In: Moccia FD, Palestino MF (eds) Planning stormwater resilient urban open spaces. CLEAN, Naples, pp 162–177

Coppola E (2016) Infrastrutture sostenibili urbane. INU Edizioni, Roma

Cortinovis C, Geneletti D (2019) A framework to explore the effects of urban planning decisions on regulating ecosystem services in cities. Ecosyst Services 38:100946. https://doi.org/10.1016/j.ecoser.2019.100946

d'Alessandro L, Realfonzo R (eds) (2018) Per una strategia di sviluppo della Città Metropolitana di Napoli. Franco Angeli, Milano

De Luca G, Moccia FD (eds) (2017) Pianificare le città metropolitane in Italia. Interpretazioni, approcci, prospettive. INU Edizioni, Roma

European Commission (2013) Building a green infrastructure for Europe. Environment. https://doi.org/10.2779/54125

Garramone V, Gissi E (2018) Proposte di evoluzione degli standard urbanistici attraverso i servizi ecosistemici. Urbanistica Informazioni 278–279:58–61

Giaimo C, Barbieri CA (2018) Paradigmi ecosistemici, piano urbanistico e città contemporanea. L'esperienza del progetto Life Sam4cp. Urbanistica 159:114–124

Haines-Young R, Postchin MB (2018) Common International Classification of Ecosystem Servicies (CICES) V 5.1 and Guidance on the Application of the Revised Structure. Available at: www.cices.eu

Indovina F (2018) Il "diritto alla città" oggi. Archivio di Studi Urbani e Regionali 122:5–9

Lefebvre H (1968) Le droit à la ville. Edition Anthropos, Parigi

Moccia FD (2009) L'urbanistica nella fase dei cambiamenti climatici. Urbanistica 140:95–102

Moccia FD (2013a) Ecological restoration methodology. In: Moccia FD, Palestino MF (eds) Planning stormwater resilient urban open spaces. CLEAN, Naples, pp 12–17

Moccia FD (2013b) Premise. Urban restoration of the stormwater system. In: Moccia FD, Palestino MF (eds) Planning stormwater resilient urban open spaces. CLEAN, Naples, pp 88–95

Moccia FD (2014) Restauro Paesaggistico ambientale. In: Petroncelli E (ed) Progetto paesaggio tra letteratura e scienza. Liguori Editore, Napoli, pp 133–144

Moccia FD (2019) Drenaggio sostenibile, infrastrutture verdi, urbanistica. Sustainable drainage, green infrastructure, urbanism. In: D'Ambrosio V, Leone MF (eds) Progettazione ambientale per l'adattamento al Climate Change Adaptation. 2. Tools and guidelines for climate risk reduction. CLEAN, Napoli, pp 226–235

Moccia FD, Berruti G (2018) Strategies of landscape restoration and city naturalizing. In: Cocci Griffoni R, D'Onofrio R, Sargolini M (eds) Quality of life in urban landscapes. In search of a decision support system. Springer, Cham, pp 365–378

Moccia FD, Scalera B (2018) La strategia di rigenerazione territoriale. Reti di connessione e valorizzazione dei luoghi d'eccellenza dei Campi Flegrei. INU Edizioni, Roma

Moccia FD, Sgobbo A (2016) Flood Hazard: planning approach to risk mitigation and periphery rehabilitation. In: Syngellakis S (ed) Management of natural disasters. WIT Press, Southampton, pp 129–144

Novotny V (2007) Effluent dominated water bodies, their reclamation and reuse to achieve sustainability. In: Novotny V, Brown P (eds) Cities of the future. IWA Publishing, London, pp 191–215

Oppio A, Bottero M, Arcidiacono A (2018) Assessing urban quality: a proposal for a MCDA evaluation framework. Ann Op Res:1–18. https://doi.org/10.1007/s10479-017-2738-2

Salzano E (2007) Fondamenti di urbanistica. La storia e la norma. La Terza Editori, Roma-Bari

Chapter 9
Green Infrastructure and Local Planning Processes: A Study Concerning the Metropolitan Context of Cagliari

Sabrina Lai, Federica Leone, and Corrado Zoppi

Abstract This study aims at defining planning policies that can enhance the strength and quality of a regional green infrastructure by improving its components. An innovative methodology that couples ecosystem services mapping, spatial analyses, and multiple linear regressions is used and applied to three municipalities belonging to the Metropolitan City of Cagliari (Italy). The results of the analysis show that a number of planning choices implied by the municipal masterplans and their zoning layouts can affect the suitability of a patch to be included in the regional green infrastructure; among such choices are the presence of conservation areas within the urban tissue, access to relevant natural or cultural assets, and improved natural and semi-natural habitats in rural areas. Limitations of the proposed methodology, as well as replicability issues and directions for future research, are also identified.

Keywords Green infrastructure · Ecosystem services · Land-use planning · Zoning schemes · Municipal masterplans · Regression models

9.1 Introduction

At the end of the 1990s, the term green infrastructure (GI) represented an emerging research topic in relation to landscape planning (Mell 2016). Thenceforth, due to its multifunctional character, the concept of GI has been associated with various disciplines, such as landscape ecology (Jongman and Pungetti 2004), green space

S. Lai
Department of Environmental Protection, Regione Autonoma della Sardegna, Cagliari, Italy

F. Leone (✉) · C. Zoppi
Department of Civil and Environmental Engineering, and Architecture, Università of Cagliari, Cagliari, Italy
e-mail: federicaleone@unica.it

A. Arcidiacono, S. Ronchi (eds.), *Ecosystem Services and Green Infrastructure*, Cities and Nature, https://doi.org/10.1007/978-3-030-54345-7_9

planning (Fábos 2004), and water resources management (Ahern 2007). Indeed, GI provides several benefits in terms of biodiversity conservation (Benedict and McMahon 2006) and of improving health and well-being of local communities and society (Kambites and Owen 2006). Several authors (Benedict and McMahon 2006; Wright 2011; Weber et al. 2006; European Commission 2013) have provided their own take on the meaning of GI. For example, Benedict and McMahon (2006) stress the socio-economic benefits connected with GI, although the definition they provide identifies GI as an ecological system aimed at supporting three components of human well-being (environmental, social, and economic). While recognizing the importance of connectivity, multifunctionality, and green spaces within the definition of GI, Wright (2011) argues that providing a deterministic definition of GI is in conflict with its flexible and continuously evolving character that makes the category of GI appropriate and effective to be applied in different research fields. Weber et al. (2006) define GI as an ecological system composed of natural and semi-natural areas focusing on the environment-related character of GI. Finally, the European Commission (2013, p. 3) provides the following generally shared definition of GI "*a strategically planned network of natural and semi-natural areas with other environmental features designed and managed to deliver a wide range of ecosystem services.*"

On the other hand, despite the undeniable benefits, the implementation of the GI concept and related principles is problematic and restricted to few forward-looking cases (Di Marino et al. 2019).

From this theoretical perspective, this study discusses the potential relationships between planning policies defined by municipal masterplans (MMPs) and GI, identified in relation to a methodology defined and applied to the Sardinian Region in a few recent studies (Cannas et al. 2018; Lai et al. 2018), where four values (natural, conservation, landscape, and recreational) are deemed to identify areal units to be included in the Sardinian green infrastructure (SGI); under this perspective, this study is based on two phases. In the first phase, the spatial configuration of areas eligible to be included in the SGI is overlaid with the zoning schemes of MMPs. In the second phase, a regression analysis is implemented in order to define and analyze correlations between the spatial configuration and MMPs' provisions. Three municipalities (Assemini, Cagliari, and Capoterra) of the Sardinian Metropolitan City of Cagliari are taken as the spatial contexts for the application of the proposed methodology. Moreover, the study proposes recommendations and suggestions in terms of planning policies, which are based on the assessment of the relationships between the above-mentioned four values and the zone types defined by the MMPs.

The proposed methodology can be easily exported to other urban contexts at national and international levels and this study may represent a significant starting point to conduct a comparative analysis between different case studies.

9.2 Materials and Methods

9.2.1 Case Study

In Italy, local governments are in charge of regulating land uses within the municipal boundaries. Therefore, they elaborate, adopt, and approve MMPs, and by doing so, they define strategies and policies concerning land-use changes (Commission of the European Communities 2000). Moreover, the Italian planning system is characterized by a hierarchical structure (Commission of the European Communities 2000, p 35) where MMPs have to be consistent with the instruments and strategies established by higher tier authorities, such as the Regional Landscape Plan (RLP).[1]

Assemini, Cagliari, and Capoterra, the three cities considered in this study, are municipalities administered by two councils[2] (*Giunta* and *Consiglio*). The *Giunta*, that is the municipal government, is chaired by the mayor. Moreover, the three municipalities belong to the Metropolitan City of Cagliari (Fig. 9.1), established in

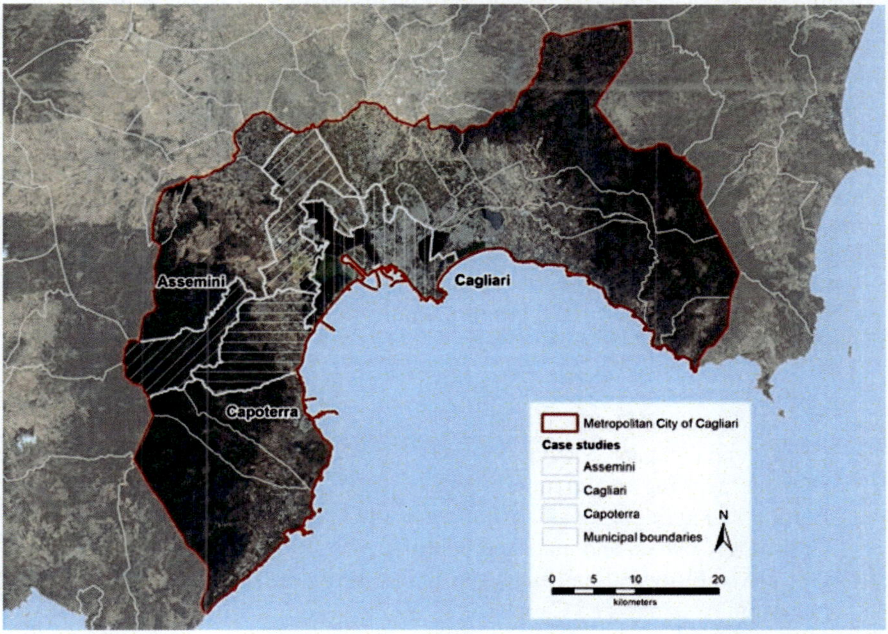

Fig. 9.1 Study area: the municipalities of Assemini, Cagliari and Capoterra within the Metropolitan City of Cagliari

[1] http://www.sardegnageoportale.it/index.php?xsl=2420&s=40&v=9&c=14482&na=1&n=10&esp=1&tb=14401 (Accessed 5 Aug 2019).

[2] The Giunta is composed of members designated by the mayor and is entitled for municipal executive functions. The Consiglio is composed of members elected by the residents and is entitled for political and administrative control functions.

compliance with the national Law no. 2014/56 and the regional Law no. 2016/2. In particular, Cagliari, the regional capital, is the most populated town of the Metropolitan City, with a population of approximately150,000 residents and a municipal area of 85 km^2. Assemini and Capoterra are located in the southwestern side of the Metropolitan City, with a population of around 27,000 and 23,000 inhabitants and a surface of 118 and 69 km^2, respectively.

Not only are the two medium-sized towns connected geographically with Cagliari but also economically. Indeed, around 30% of Assemini's and Capoterra's residents commute to work from the hinterland to Cagliari (ISTAT 2019).

9.2.2 *Zoning Schemes*

A recently approved MMP is available for both Assemini and Capoterra. Such plans, approved in 2015 and 2016 respectively, are compliant with the Sardinian RLP and are available through the municipalities' websites.[3] On the contrary, in Cagliari a 15-year-old plan, approved prior to the RLP, is in force; therefore, the plan is yet not compliant with the RLP and the conflictual adjustment process (Zoppi and Lai 2010) has yet to start. As with Capoterra and Assemini, Cagliari's planning documents can be retrieved from its institutional website.[4]

The three zoning schemes, one for each municipality, were analyzed through the lenses of their planning implementation codes. In order to reduce the variety of zone types, the zoning schemes (which are similar, since all of them are compliant with the same regional normative framework, which provides a taxonomy for zone types, although differences in subtypes are possible) were simplified. For example, subtypes of the same zone type were joined, or areas sharing similar building codes were merged. As a result, 10 types of planning zones were identified, which are as follows:

- A: historic areas
- B: residential completion zones
- C: residential expansion zones
- D: manufacturing and commerce areas
- E: rural and agricultural areas (not present in Cagliari)
- G: privately owned and managed public services
- GS: privately owned and managed public services, subtype: urban parks (not present in Assemini and Capoterra)
- H: conservation, protection, and buffer areas

[3] https://comune.assemini.ca.it/amministrazione/amministrazione-trasparente/pianificazione-governo-del-territorio/piani-programmi-16; https://www.comune.capoterra.ca.it/index.php?option=com_content&view=article&id=2817&Itemid=532 (Accessed 5 Aug 2019).

[4] https://www.comune.cagliari.it/portale/it/at18_puc.page; https://sit.comune.cagliari.it/?filtro=puc#13/39.2238/9.0906 (Accessed 5 Aug 2019).

- IC: mixed-use areas (residential and service areas) (not present in Assemini and Capoterra)
- S: publicly owned public services (e.g., small parks or parking lots).

9.2.3 Methodology

In this study, we apply the methodology developed by Arcidiacono et al. (2016) to map a potential regional GI in Lombardy (Italy), slightly modified following a set of articles focusing on Sardinia (Lai and Leone 2017; Cannas et al. 2018; Lai et al. 2018), where the authors assess the suitability for a patch to be included in a regional GI on the basis of the following elements:

- Natural value (V_NAT), which accounts for biodiversity richness and quality despite anthropogenic threats, hence a proxy for biodiversity's capacity to provide ecosystem services
- Conservation value (V_CON), accounting for high-quality natural and semi-natural habitats enlisted in the European Union (EU), which are the core of a GI according to the European Commission (2013)
- Recreation value (V_REC), expressing landscapes' recreation attractiveness through an estimation of visitation data
- Landscape value (V_LAND), which assesses landscapes' scenic quality through an assessment of the legal protection framework implied by the RLP.

Each of the above-listed elements accounts for one of the several functions provided by a GI. Once the four elements are separately mapped and assessed (each normalized in the 0–1 range), the suitability of each patch to belong to an SGI is expressed as the sum of the four values (i.e., Total value, T_VAL) in that patch. The T_VAL map (Fig. 9.2) therefore represents the suitability map, or, in other words, the potential SGI.

Through a spatial intersection between the T_VAL map and the zoning layouts of the MMPs (whose only attribute is "Zone," expressing in each point, the zoning type allocated by the MMP, Fig. 9.3), a new vector layer is generated, where for each polygon, the attributes of the two input layers are joined (V_NAT, V_CON, V_REC, V_LAND, T_VAL, and Zone). A multiple linear regression is next performed separately for each of the three municipalities using the attribute table of the output layer from the spatial intersection:

$$\mathbf{TotVal}_{\mathrm{k}} = \beta_{0,\mathrm{k}} + \beta_{1,\mathrm{k}}\mathrm{A} + \beta_{2,\mathrm{k}}\mathrm{B} + \beta_{3,\mathrm{k}}\mathrm{C} + \beta_{4,\mathrm{k}}\mathrm{D} + \beta_{5,\mathrm{k}}\mathrm{E} + \beta_{6,\mathrm{k}}\mathrm{G} + \beta_{7,\mathrm{k}}\mathrm{GS} + \beta_{8,\mathrm{k}}\mathrm{H} + \beta_{9,\mathrm{k}}\mathrm{IC} + \beta_{10,\mathrm{k}}\mathrm{Area}$$

where:

k (= 1, 2, or 3) is a municipality.

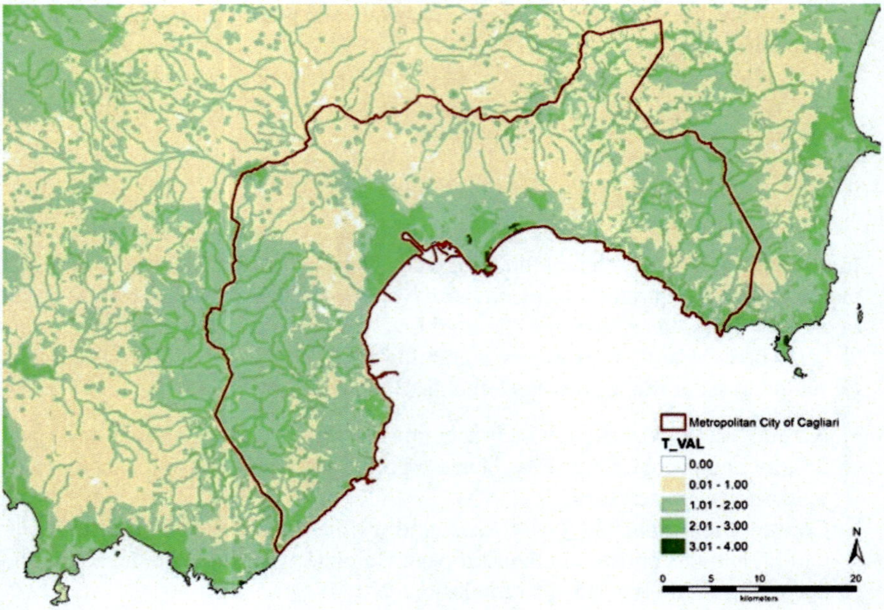

Fig. 9.2 Map of the total value (T_VAL) in the study area

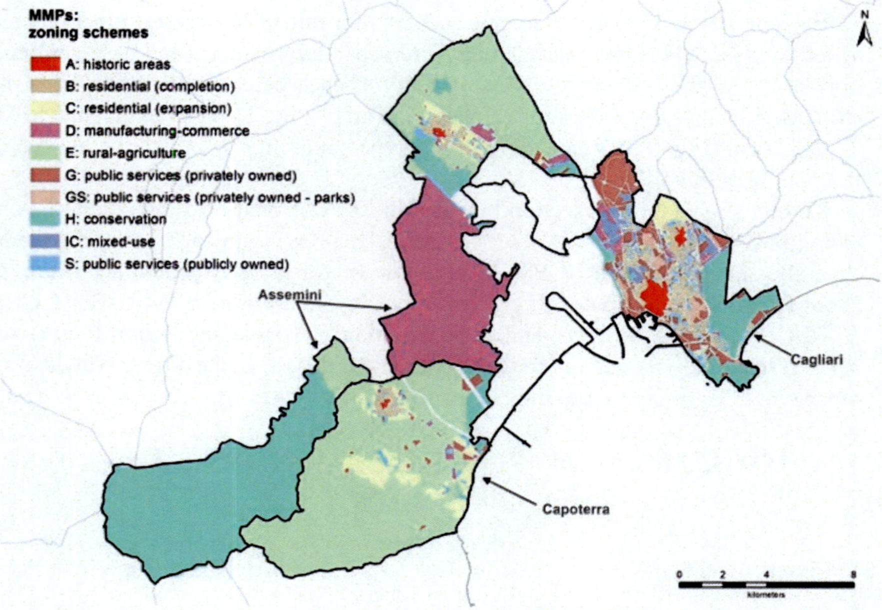

Fig. 9.3 Zoning schemes (as per the municipal masterplans) in the study area

A, *B*, …, *IC* are dichotomous explanatory variables accounting for the zone types; each can only take two values (0 or 1), as follows: if a patch belongs to the A zone type, then A = 1, otherwise A = 0; if a patch belongs to the B zone type, then B = 1, otherwise B = 0; likewise for any of the remaining zone types included in the regression (C to IC).

β_i (i = 1, …, 9) is a coefficient, estimated through the regression, that identifies the change in T_VAL for the corresponding zone type (e.g., β_1 for A, β_2 for B, and so on). Such change is therefore estimated in relative terms, by taking as basic condition the patch's allocation to the S zone. Therefore, a taxonomy of the zone types can be defined based on the β_i coefficients or, more precisely, on the quantitative contributions to T_VAL expressed by β_i's values.

Area is the size of a polygon in the output layer.

Finally, for each municipality, an ordered list of the zone types is developed based on the regression's results, where the order depends on the value of the β_i coefficients.

9.3 Findings

The regression results concerning Capoterra, Assemini, and Cagliari show the impacts of the zone type of a patch on its qualification to be part of the SGI. The estimated coefficients of variables labeled "A zones" through "H zones", which include "IC zones" and "GS zones" in the case of Cagliari, where the "E zones"-labeled variable does not appear, identify the impacts on the qualification of a patch to be part of the SGI with respect to the reference condition of a patch being located in an "S zone" (Tables 9.1, 9.2, and 9.3). These impacts equal, ceteris paribus, the differentials in T_VAL. The zones classed A through H can be ranked in the same way, based on the quantitative size of the impacts.

The ranking of the zones as regards their eligibility to be included in the SGI is the following:

1. Areas characterized by the presence of historic heritage related to urban fabrics and buildings (A zones) except for Capoterra, whose historic center is less important than Assemini's and Cagliari's analogous zones
2. Rural and agricultural areas (E zones, which are not included in the Cagliari's zoning code), featured by almost irrelevant land-taking processes
3. Protection and conservation areas (H zones), featured by spatial units characterized by relevant landscape and environmental values.

The impact of collective services (GS zones), characterized by the presence of leisure areas and public parks, that is, areas mainly consisting of open spaces (a zone type which is only included in the Cagliari's zoning code), is in line with the impact of the rural and agricultural areas (H zones).

Table 9.1 The homogeneous zones of the zoning code of the municipal masterplan of Cagliari ranked on the basis of their impact on T_VAL entailed by the regression estimates (rank (R) from 1 to 7) and mean values of V_NAT, V_CON, V_LAND, and V_REC detected as regards each zone type (EX: the zone does not show up in the zoning code; Nul: the estimate shows a non-significant p-value)

		Cagliari			
		Mean values			
Class (zoning code)	R	V_NAT	V_CON	V_LAND	V_REC
Historic areas (A zones)	2	0.429	0.000	1.000	0.568
Residential completion areas (B zones)	6	0.029	0.000	1.000	0.259
Residential expansion areas (C zones)	Nul				
Manufacturing and commerce areas (D zones)	7	0.229	0.000	0.948	0.040
Rural and agricultural areas (E zones)	EX				
Privately owned and managed public services (G zones)	Nul				
Privately owned and managed public services (GS zones)	3	0.610	0.021	1.000	0.259
Conservation, protection and buffer areas (non-building areas, H zones)	1	0.676	0.200	1.000	0.192
Mixed-use areas (residential and service areas, IC zones)	4	0.242	0.000	0.987	0.100
Publicly owned public services (S zones)	5	0.099	0.001	1.000	0.223

Table 9.2 The homogeneous zones of the zoning code of the municipal masterplan of Assemini ranked on the basis of their impact on T_VAL entailed by the regression estimates (rank (R) from 1 to 6) and mean values of V_NAT, V_CON, V_LAND and V_REC detected as regards each zone type (EX: the zone does not show up in the zoning code; Nul: the estimate shows a non-significant p-value)

		Assemini			
		Mean values			
Class (zoning code)	R	V_NAT	V_CON	V_LAND	V_REC
Historic areas (A zones)	4	0.000	0.000	0.719	0.090
Residential completion areas (B zones)	6	0.040	0.000	0.199	0.063
Residential expansion areas (C zones)	Nul				
Manufacturing and commerce areas (D zones)	2	0.642	0.159	0.638	0.009
Rural and agricultural areas (E zones)	3	0.479	0.030	0.349	0.005
Privately owned and managed public services (G zones)	Nul				
Privately owned and managed public services (GS zones)	EX				
Conservation, protection and buffer areas (non-building areas, H zones)	1	0.750	0.190	0.650	0.006
Mixed-use areas (residential and service areas, IC zones)	EX				
Publicly owned public services (S zones)	5	0.319	0.000	0.260	0.029

Table 9.3 The homogeneous zones of the zoning code of the municipal masterplan of Capoterra ranked on the basis of their impact on T_VAL entailed by the regression estimates (rank (R) from 1 to 6) and mean values of V_NAT, V_CON, V_LAND, and V_REC detected as regards each zone type (EX: the zone does not show up in the zoning code; Nul: the estimate shows a non-significant p-value)

		Capoterra			
		Mean values			
Class (zoning code)	R	V_NAT	V_CON	V_LAND	V_REC
Historic areas (A zones)	Nul				
Residential completion areas (B zones)	5	0.050	0.000	0.030	0.070
Residential expansion areas (C zones)	Nul				
Manufacturing and commerce areas (D zones)	Nul	Nul	Nul	Nul	Nul
Rural and agricultural areas (E zones)	3	0.530	0.059	0.519	0.009
Privately owned and managed public services (G zones)	2	0.450	0.060	0.640	0.020
Privately owned and managed public services (GS zones)	EX				
Conservation, protection and buffer areas (non-building areas, H zones)	1	0.700	0.279	1.000	0.040
Mixed-use areas (residential and service areas, IC zones)	EX				
Publicly owned public services (S zones)	4	0.290	0.019	0.509	0.033

The taxonomy of the zones is shown in Tables 9.1, 9.2 and 9.3. The taxonomy represents the ranking of the zones with reference to the size of their impact on the qualification to be included in the SGI and the average values of V_NAT, V_CON, V_LAND, and V_REC, which make it possible to compare the covariates' impacts.

9.4 Concluding Remarks

The zoning codes of three towns of the Metropolitan City of Cagliari are analyzed in order to detect correlations between zone types and the eligibility of spatial units to be part of the SGI, determined through the methodological approach of a previous study (Lai and Leone 2017).

The findings reported in the third section indicate that the non-building zones identified as H areas always generate positive impacts on the qualification of spatial units to be included in the SGI. Moreover, the mean values of the determinants of T_VAL are often lower than 1, which is the maximum value they can take, and sometimes they are lower than 0.5. This implies that there is still room for enhancing the performance of the spatial units of the three towns as regards their eligibility to be part of the SGI by exploiting the capacity to increase the factors whose values are more or less far from 1, namely:

1. V_NAT is always higher than 0.5
2. V_CON and V_REC are always lower than 0.5
3. V_LAND equals 1 in Capoterra and Cagliari, whereas it equals 0.650 in Assemini.

That being so, the performance can improve not only by focusing on V_CON and V_REC, but also by relying on the increase of V_NAT.

V_CON depends on the size of areas identified as community interest habitats. The non-building zones (H zones) shelter areas characterized by the presence of valuable natural resources, among which are the sites of the Natura 2000 Network or the ecological corridors used by migratory species. A policy implication of the results of the study is that the conservative approach, which features the Natura 2000 Network, should be exported outside the Network's sites so as to increase the eligibility of external areas to be included in the SGI. The expansion of conservation measures would be possible provided that habitats of community interest were identified outside the Natura 2000 sites, which would imply an increase in the spatial analysis concerning the identification of habitats of community interest. Applied scientific research and public consensus would be relevant preconditions to increase the size of the H conservation, protection, and buffer areas. This would also entail that planning approaches which implement protection-oriented spatial policies into protected areas and their surroundings should be encouraged in order to enhance the impacts of conservation measures not only within Natura 2000 sites but also in their buffer areas (Maiorano et al. 2007).

With reference to V_REC, the study's findings put in evidence that there is more potential for enhancement than in the case of V_CON, since its average values are always lower as regards the non-building H areas. V_REC is based on information drawn from Flickr and it shows a location's attractiveness in a given time span. A number of articles argue that a location's leisure- and tourism-related attractiveness depends on the endowment of features such as accommodation facilities and transport infrastructure (Font 2000; Amoako-Tuffour and Martinez-Espineira 2012; Heagney et al. 2018). As a consequence, our findings imply that the issue of accessibility is central. This entails that new paths for bikers and walkers would be needed, which should be carefully designed in order to avoid increase in landscape fragmentation. Indeed, a fragmentation increase would imply a decrease in V_NAT and V_CON.

Even though V_NAT takes mean values included in the 0.66–0.74 interval, some enhancement is still possible. V_NAT suffers from improper uses of land and from human activities, which threaten habitats included in the Natura 2000 sites and listed in their standard data forms. Under this perspective, implications for planning policies are pretty straightforward and consist of control and limitation of land-take and habitat-threatening operations.

Furthermore, since the presence of the H conservation, protection, and buffer areas has a positive impact on the suitability of a spatial unit to be included in the SGI, the findings entail a pledge on behalf of planning policies for an increase in the present size of the H areas and for the identification of new such areas which should be substitute for other types of zones.

As for Assemini and Capoterra, the E rural and agricultural areas show a positive impact on spatial units' suitability to be included in the SGI. V_CON's, V_NAT's, V_REC's, and V_LAND's highlight lower mean values with respect to the H areas, and, that being so, spatial policies can lead to an increase in the eligibility of the patches located in the E areas to be included in the SGI, especially as regards V_CON and V_NAT. The qualitative status of land cover is the most important factor that determines V_NAT, which is often threatened by fragmentation of rural landscapes generated by infrastructure development and by intensive farming. According to He et al. (2017), the quality of habitats is significantly influenced by land covers, and it can be enhanced through the implementation of land uses, which should focus on small and isolated agricultural villages. As for V_CON, measures aimed at integrating agriculture, climate, and environment are planned by the Rural Development Programme concerning the Sardinian Region, which includes financial support to the relocation of shares of the agricultural land to wilderness, such as development of natural grass buffer strips along lagoons, identification of areas bound to remain uncultivated in order to conserve and improve the status of wild animals and plants, or enhancing the protection of small mammals and birds by strengthening natural safe paths such as the ecological corridors of the Natura 2000 Network.

Finally, it has to be stressed that the methodological approach proposed in this study is easily exportable to other spatial contexts of the EU, since Natura 2000 sites are established across all the countries of the EU, under the provisions of the Habitats Directive. Indeed, the implementation of the proposed methodology in other contexts may possibly imply the use of different sets of explanatory variables in the regression models in order to fit the characteristics of different spatial contexts, which can be straightforwardly implemented.

Two problematic questions have to be highlighted as directions for future research. First, measuring the size of attractiveness (V_REC) exclusively on the basis of data made available by social media is somewhat questionable, even though a number of studies put in evidence that this kind of information is highly reliable in order to analyze the factors which characterize the visitors' behaviors and attitudes (Wood et al. 2013; Sessions et al. 2016; Heikinheimo et al. 2017). Secondly, the assessment of V_NAT does not take into consideration nature-based solutions, such as detention basins, permeable parking areas, and floodable zones. This is likely to lead to undervaluing V_NAT in highly urbanized areas.

Acknowledgments This paper is written within the Research Program "Rural landscapes of Sardinia: Spatial planning concerning blue and green infrastructure and complex networks," funded by the Autonomous Region of Sardinia for the period 2019–2021, under the provisions of the Call for the presentation of "Projects related to fundamental or basic research" of the year 2017, implemented at the Department of Civil and Environmental Engineering and Architecture (DICAAR) of the University of Cagliari, Italy.

Sabrina Lai, Federica Leone, and Corrado Zoppi have jointly made substantial contributions to the study's conception, design, and concluding remarks (Sect. 9.4). Sabrina Lai has taken care of Sect. 9.1 (Introduction); Federica Leone has taken care of Sect. 9.2 (Materials and Methods); and Corrado Zoppi has taken care of Sect. 9.3 (Findings).

References

Ahern J (2007) Planning and design for sustainable and resilient cities: theories, strategies and best practice for green infrastructure. In: Novotny V et al (eds) Water-centric sustainable communities. Wiley-Blackwell, Hoboken, pp 135–176

Amoako-Tuffour J, Martinez-Espineira R (2012) Leisure and the net opportunity cost of travel time in recreation demand analysis: an application to Gros Morne National Park. J Appl Econ 15:25–49

Arcidiacono A, Ronchi S, Salata S (2016) Managing multiple ecosystem services for landscape conservation: a green infrastructure in Lombardy region. Proc Eng 161:2297–2303

Benedict MA, McMahon E (2006) Green infrastructure: linking landscapes and communities. Island Press, Washington, DC

Cannas I, Lai S, Leone F, Zoppi C (2018) Green infrastructure and ecological corridors: a regional study concerning Sardinia. Sustainability 10:1265

Commission of the European Communities (CEC) (2000) The EU compendium of spatial planning systems and policies: Italy. Office for Official Publications of the European Communities, Luxembourg

Di Marino M, Tiitu M, Lapintie K, Viinikka A, Kopperoinen L (2019) Integrating green infrastructure and ecosystem services in land use planning. Results from two Finnish case studies. Land Use Policy 82:643–656

European Commission (2013) Communication from the Commission to the European Parliament, the Council, the European Economic and Social Committee and the Committee of the Regions. Green infrastructure (GI) — enhancing Europe's natural capital. European Commission, Brussels. Available online. https://eur-lex.europa.eu/resource.html?uri=cellar:d41348f2-01d5-4abe-b817-4c73e6f1b2df.0014.03/DOC_1&format=PDF. Accessed 5 Aug 2019

Fábos JG (2004) Greenway planning in the United States: its origins and recent case studies. Landsc Urban Plan 68:321–342

Font AR (2000) Mass tourism and the demand for protected natural areas: a travel cost approach. J Environ Econ Manag 39:97–116

He J, Huanga J, Li C (2017) The evaluation for the impact of land use change on habitat quality: a joint contribution of cellular automata scenario simulation and habitat quality assessment model. Ecol Model 366:58–67

Heagney EC, Rose JM, Ardeshiri A, Kovac M (2018) Optimising recreation services from protected areas — understanding the role of natural values, built infrastructure and contextual factors. Ecosyst Serv 31:358–370

Heikinheimo V, Di Minin E, Tenkanen H, Hausmann A, Erkkonen J, Toivonen T (2017) User-generated geographic information for visitor monitoring in a national park: a comparison of social media data and visitor survey. ISPRS Int J Geo-Inf 6:85

ISTAT (2019) Censimento Popolazione Abitazioni [National Census]. Available online. http://dati-censimentopopolazione.istat.it/Index.aspx#. Accessed 5 Aug 2019

Jongman R, Pungetti G (2004) Ecological networks and greenways: concept, design and implementation. Cambridge University Press, Cambridge

Kambites C, Owen S (2006) Renewed prospects for green infrastructure planning in the UK. Plann Pract Res 21:483–496

Lai S, Leone F (2017) A methodological approach to identify a multifunctional green infrastructure at the regional scale. A case study from Sardinia, Italy. Urbanistica Informazioni 272:836–840
Lai S, Leone F, Zoppi C (2018) Implementing green infrastructures beyond protected areas. Sustainability 10:3544
Maiorano L, Falcucci A, Garton EO, Boitani L (2007) Contribution of the Natura 2000 Network to biodiversity conservation in Italy. Conserv Biol 21:1433–1444
Mell IC (2016) Global green infrastructure: lessons for successful policy-making, investment and management. Routledge, Abingdon
Sessions C, Wood SA, Rabotyagov S, Fisher DM (2016) Measuring recreational visitation at U.S. national parks with crowd-sourced photographs. J Environ Manag 183:703–711
Weber T, Sloan A, Wolf J (2006) Maryland's green infrastructure assessment: development of a comprehensive approach to land conservation. Landsc Urban Plan 77:94–110
Wood SA, Guerry AD, Silver JM, Lacayo M (2013) Using social media to quantify nature-based tourism and recreation. Sci Rep-UK 3:2976
Wright H (2011) Understanding green infrastructure: the development of a contested concept in England. Local Environ 16:1003–1019
Zoppi C, Lai S (2010) Assessment of the regional landscape plan of Sardinia (Italy): a participatory-action-research case study type. Land Use Policy 27:690–705

Chapter 10
Ecosystem Services Integration into Local Policies and Strategies in the City of Bologna: Analysis of the State of the Art and Recommendations for Future Development

Claudia de Luca, John Martin, and Simona Tondelli

Abstract The enhancement of the role of ecosystem services to guarantee the sustainable transformation of urban areas has been at the centre of sustainable development discussion for some years. Nevertheless, the integration of such concepts into local policies and strategies is still far from complete. In this chapter, first the issue of defining and classifying relevant ecosystem services for urban areas is presented, followed by a brief presentation of the case study analysed.

The chapter presents an analysis of the integration of urban ecosystem services (UES) concepts into the City of Bologna's urban planning, greening and climate policies. Even though the overall framework of UES is not explicitly mentioned in Bologna's urban policies, as only the Climate Adaptation Plan clearly refers to it, several UES – mostly regulating and cultural services – have been included in the documents analysed including indicators, criteria and parameters to enhance them. However, an overall framework with clear targets and monitoring programme is currently missing, leaving great opportunities for the integration of UES concepts within the new City Master Plan.

Keywords Ecosystem services · Urban planning · Urban policies · Bologna · Nature-based-solutions · Sustainable transition

---

C. de Luca (✉) · S. Tondelli
Department of Architecture, Alma Mater Studiorum University of Bologna, Bologna, Italy
e-mail: claudia.deluca5@unibo.it

J. Martin
University of Plymouth, Plymouth, UK

A. Arcidiacono, S. Ronchi (eds.), *Ecosystem Services and Green Infrastructure*, Cities and Nature, https://doi.org/10.1007/978-3-030-54345-7_10

10.1 Introduction

According to the UN, 68% of the world population is expected to live in cities by 2050 (Nations et al. 2018) acknowledging the constant migration flow from rural to urban areas that characterised the last century. At the same time, climate change and environmental pressures have become an urgent threat that cannot be postponed.

Ensuring sustainable, resilient and healthy quality of life to citizens will be among the most important priorities for local governments. Services provided by urban ecosystems, such as climate regulation, water purification, surface water run-off control, cultural recreation, food provisioning, etc., could strongly support urban areas in creating healthier and more sustainable environments. Several cities, such as Barcelona (Baró et al. 2016), have already started to work on their transformation through the integration of urban ecosystem services (UES) into sustainable urban planning. The provision of such services depends on the urban ecosystem and on the quantity, quality and availability of green areas within the city area. Planning and environmental disciplines are recognising the crucial role of UES (Hansen et al. 2015; Kabisch 2015; Kaczorowska et al. 2016); however, the gap between research and practice is still varied. The integration of the UES concept into local plans and its translation into actions and practices is still fuzzy. In recent years, the academic community focused on this topic (Kaczorowska et al. 2016; Woodruff and BenDor 2016; Cortinovis and Geneletti 2018) and the findings are contributing to a better understanding of what is still needed to improve UES integration into urban plans and policies. However, there is still fragmented and limited knowledge on the topic.

Bologna is an interesting case for studying this integration. It was the first Italian city to develop the Climate Adaptation Plan and it has been a forerunner in including environmental indicators and criteria into urban plans. Moreover, Bologna City Council, together with the Ministry of Economic Development of Poland, is the coordinator of the Urban Agenda partnership on Sustainable Land Use and Nature-Based Solutions (NBSs).[1]

The Urban Agenda for the EU was launched in May 2016 with the Pact of Amsterdam. The partnership on Sustainable Land Use and NBSs started in June 2017 and it includes different services from the EU Commission, 8 European cities and metropolitan urban areas, 6 National Ministries and other relevant stakeholders. The partnership agreed that the balance between urban compactness and achieving high standards of quality of life is one of the major challenges for Europe's urban areas. In this sense, the role of ecosystem services (ES) and nature-based solutions (NBSs) integration into planning documents has been part of the discussion in the development and the implementation of the Action Plan.

Based on previous research (Wilkinson et al. 2013; Hansen et al. 2015; Kabisch 2015), this chapter uses a discursive approach to understand the degree of integration of UES into Bologna's urban and environmental plans and policies considering explicit and implicit references to UES framework, categories and

[1] https://ec.europa.eu/futurium/en/sustainable-land-use

functions. The primary questions guiding this analysis are (a) To what degree are UES integrated into Bologna's urban plans, strategies and technical guidelines? (b) How effective is such integration in terms of actual implementation, monitoring and impact assessment? (c) Based on this analysis, how can the gap between science and practices be bridged so that future master plans, policies and technical guidance are more effective?

This contribution attempts to reply to the questions above by first presenting a selection of UES pertinent for Bologna and UES relevant local plans, strategies and policies to be analysed (Sect. 10.2), followed by a description of the method used for the analysis (Sect. 10.2). The main findings of the policy analysis will be presented in Sect. 10.3, while Sect. 10.4 will present a discussion on the degree of inclusion of UES focusing on recommendations for a better integration and effectiveness of such concepts.

10.2 Material and Methods

10.2.1 Case-Study Presentation

According to the statistical department of the municipality of Bologna, 390.636 people lived in the city at the end of 2018 (+0,4% compared to 2017). Bologna has slowly grown in the last years, due to an increasing immigration, mostly from other Italian regions. The city covers an area of approximately 140 square km and is characterised by a humid subtropical climate (Cfa Koppen classification). The city has seen the occurrence of severe droughts and heat waves increase. Such events are only likely to increase further as climate projection for the time period 2021–2050 foreseen a 2° average temperature increase (Bologna Municipality 2015b). For this reason, in 2015, Bologna adopted a Climate Adaptation Plan and introduced greening and environmental measures into several other relevant urban and climate policies and strategies.

10.2.2 Methodology for the Analysis

To understand the degree of inclusion of urban ecosystem services (UES) into Bologna's urban, climate and greening policies and strategies, a three-step process has been followed:

- Definition and selection of UES pertinent for the Bologna case
- Selection of local policies, plan and strategies potentially relevant for UES
- Qualitative content analysis of selected local policies, plan and strategies

10.2.2.1 Definition and Selection of UES

The concept of cities as complex socio-ecological systems (Mascarenhas et al. 2015; Frank 2017) frames the idea of urban areas as hot-spots of human and nature networks, interactions and relations. As introduced by the Millennium Ecosystem Assessment (ME Assessment 2005), the ecosystem services framework aimed at defining and quantifying those interactions and relations as benefits or trade-off that people obtain from ecosystems. Narrowing this framework at the city level, this study considers the following urban ecosystems: street trees, lawns/parks, urban forests, cultivated land, wetlands, lakes/sea and streams (Bolund and Hunhammar 1999). From the 17 groups of services listed by Costanza et al. (1997), 12 groups of the four categories were selected, as having a major importance in Bologna areas:

- Regulating services: air filtering (gas regulation), micro-climate regulation, noise reduction (disturbance regulation), run-off control and water purification (water regulation), pollination
- Supporting services: habitat for species (refugia), genetic resources
- Provisioning services: food production and fresh water (water supply)
- Cultural services: recreational and cultural values (spiritual and educational services)

Other services such as erosion control and waste treatment could also be relevant in urban areas but are not considered significant for the case of Bologna.

10.2.2.2 Definition and Selection of Relevant Policies to Be Analysed

A varied range of urban policies can directly or indirectly affect UES provisions. For this reason, in this analysis a wide range of documents from urban planning to climate and greening policies were considered.

Urban Planning Tools

Bologna has a long and strong tradition in urban planning, and according to the Emilia-Romagna town planning law n. 20/2000, urban planning is currently addressed by 3 main tools:

- Municipal Structural Plan (Bologna Municipality 2007a – PSC): The plan was drafted in 2007 and fixes the strategy, limits and conditions to proposed urban changes. The PSC can be considered a City Master Plan: it doesn't define the exact location of the new developments or of the volumes that can be employed to foster densification and renewal policies in the well-established urban fabric. It includes a Strategic Environmental Assessment of its strategy.
- Municipal Operative Plan (Bologna Municipality 2009b – POC): The plan assigns building capacity to the areas subjected to new developments and urban

renewal. POC is a thematic document addressing only some parts of the city or some issues that are considered currently relevant.
- Building and Urban Code (Bologna Municipality 2018b – RUE): The code defines rules for building interventions and assigns specific volumes to the defined areas, in respect of the limits and conditions defined by the PSC. The RUE defines criteria for environmental performances and proposes rewards in terms of volume incentives in case of excellent standards reached.

Climate Policies

The City of Bologna recognised the importance of taking actions in the field of sustainable development in 1996, by signing the Aalborg Charter and developing the local Agenda 21 (Zanon and Verones 2013). In 2008 the city joined the Covenant of Mayors initiative, the world's largest movement for local climate and energy actions. In terms of local policies:

- The Action Plan for Sustainable Energy (Bologna Municipality 2012, SEAP), which defines actions to increase energy efficiency and use of renewable energy sources in the urban and industrial areas, focusing on construction industry, service sector, local production of energy, mobility and public facilities.
- The Climate Adaptation Plan (Bologna Municipality 2015b, CAP), the first to be approved in Italy, defines targets, actions and monitoring to support climate adaptation of the city.
- The Sustainable Urban Mobility Plan (Bologna Municipality 2018a, SUMP) that introduces important measures in terms of slow and green mobility in the city.

Environmental, Greening and Other Relevant Policies

In the context of this study, environmental and greening policies refer to strategic and technical documents regarding development, management and maintenance of green areas, such as the following:

- Municipal code for public and private green areas (Bologna Municipality 2014b), which defines the overall regulations to protect and maintain urban green areas (parks, street trees, etc.).
- Guidelines for green public areas development (Bologna Municipality 2009a), which contains procedures and recommendations to develop new public green spaces.
- Nature for children (Bologna Municipality 2015a), as a specific document which supports the development or regeneration of existing green areas in schools and kindergartens.
- Urban farming and orchards code (Bologna Municipality 2009c), which defines, promotes and regulates public orchards in the city.

- Regulation on Public Collaboration for the Urban Commons (Bologna Municipality 2014a) is the pact of collaboration, through which the city and citizens (informal groups, NGOs, private entities) agree on an intervention of care and regeneration of urban commons (green space, abandoned buildings, squares). The document is mostly relevant for the governance and the management structure of the public spaces. It allowed 280 agreements to be defined since it was developed in 2011. It is considered a pioneering document in this sense.

10.2.2.3 Qualitative Content Analysis

The qualitative analysis started with a screening of all the urban, climate and greening policies and strategies highlighted above, in order to select the most relevant documents in terms of UES enhancement and implementation.

The second step of the analysis looked for explicit reference of the overall UES framework within the analysed documents; being just one the case where the framework is explicitly mentioned the following step was to look closer at the single UES mentioned in the documents.

The third step was to classify the UES references in terms of the type of actions and interventions (awareness-raising, infrastructure, normative/regulation, financial, recommendations). This final step also checked for the relative monitoring programmes, criteria and requested performances.

10.3 Results

In total, 14 documents were screened, 10 of which were considered highly relevant, 2 potentially relevant and 2 low of relevance. All highly and potentially relevant documents were subjected to further deeper analysis (Fig. 10.1).

10.3.1 Overall References to UES

UES framework is only explicitly referenced in the Climate Adaptation Plan (Bologna Municipality 2015b). However, all the documents considered medium–highly relevant to the topic make some reference to the provision of such services from urban green areas. Fifty-eight relevant mentions have been found, referring to all the 4 UES categories and covering a wide range of potential actions and recommendations. The most recurrent category concerns regulating services (32) – generally related to micro-climate regulation and water run-off control, followed by cultural services (19) in terms of recreational activities. Provisioning and supporting services are mentioned just 7 times, generally referring to food production and habitat for species.

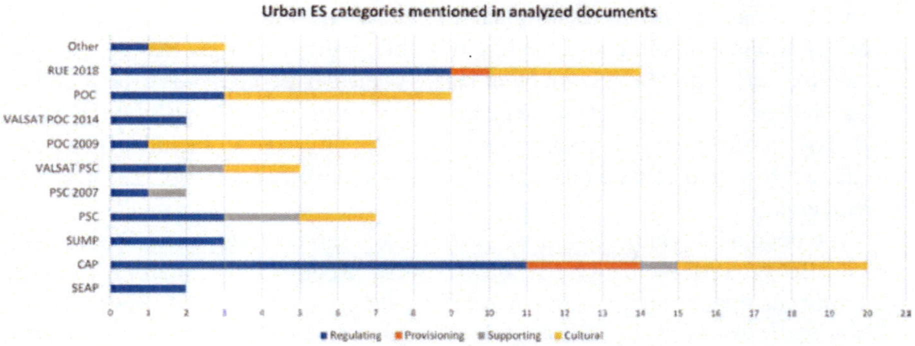

Fig. 10.1 Overall and detailed summary of UES mentions in analysed documents

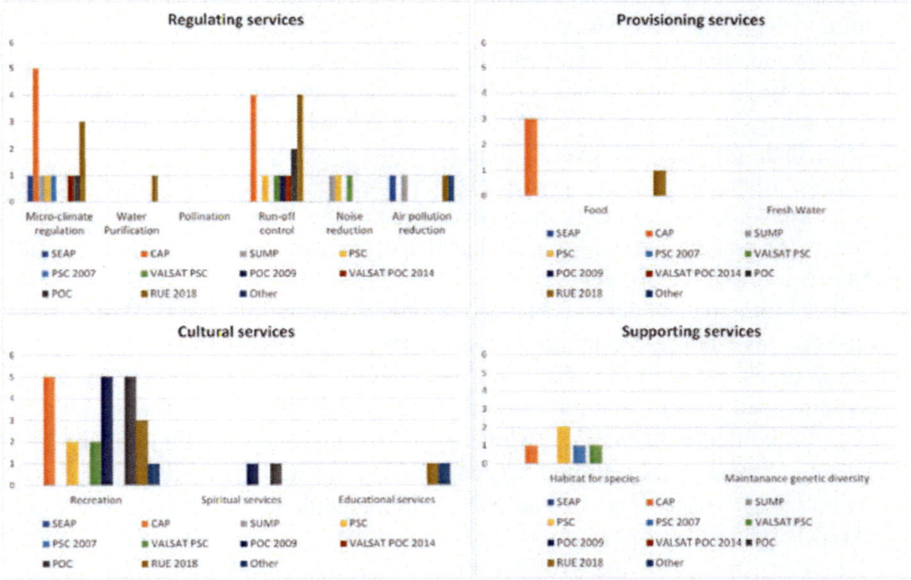

Fig. 10.2 Detailed results of UES categories per policy document

10.3.2 Detailed Results Per UES and Policy Document (Fig. 10.2)

The Climate Adaptation Plan (CAP) contains 20 references to UES, which includes the 4 UES categories, with regulating and cultural being the most frequently found. These two services are included in all the strategic lines of the document – (a) droughts and water scarcity, (b) heat waves, (c) extreme weather events and hydro-geological risk. It includes various actions for improving permeability through green areas, increasing accessibility to green public spaces for recreation and tourism, enhancing urban farming, increasing the number of trees to face air pollution, improving greening and shading of public areas, etc. The CAP also proposes guidelines to integrate UES into other local policies. It recommends to better include clear procedures on the implementation of Sustainable Urban Drainage System (SUDS) within the urban planning tools (PSC, POC) and to include trees species better adaptable to climate change within the municipal code for public and private green areas.

The document which ranked below CAP was the Building and Urban Planning code (Bologna Municipality 2018b, RUE) that proposes criteria and indicators to monitor the quality and performances of buildings, streets, public green areas, and sewerage networks. All UES categories were mentioned except supporting services. This document proposes and identifies NBSs to be implemented in urban areas and buildings such as SUDS, green roofs, buffer areas to increase vegetation and to improve accessibility and connectivity among existing public green areas to ensure multiple uses. Many of these concepts are introduced in terms of recommendations, but there are also some legally binding criteria. Developers planning new builds must comply with a minimum percentage of soil permeability metrics and contribute to micro-climate regulation by including SUDS, green roofs, green walls, etc. within their project. The overall impact of the proposed development project in terms of microclimate and permeability is then calculated through a dedicated index, known as the building impact reduction index. However, the index is calculated at the project planning phase, but not monitored during the implementation and execution of the project.

The Municipal Structural Plan (Bologna Municipality 2007a, PSC) and its Strategic Environmental Assessment (Bologna Municipality 2007b, SEA) highlight UES 7 times, twice in the plan itself and 5 times in the SEA. The plan provides recommendations in the definition of 3 ecological networks within the city and refers to the dedicated municipal operational plan (Bologna Municipality 2009b, POC) for further specifications. No concrete actions are mentioned, but the overall role of green spaces and networks in relation with microclimate regulation is acknowledged.

In compliance with the SEA Directive (Directive 2001/42/EC), the municipal structural plan has been subjected to SEA (Bologna Municipality 2009b) that provides general recommendations on the defined environmental components and on the overall objectives of the Master Plan. The identified environmental components to be analysed are: water, soil, electromagnetism, energy, mobility, air, noise and habitats for species. In the analysis of the different components, the development of new green areas to maintain and improve habitats for species and recreational activities is mentioned several times. In relation to reducing noise pollution, green barriers have been mentioned, but not considered as primary solutions to the issue. As for the soil component, SUDS are mentioned to increase soil permeability. Generally, the SEA proposed technological solutions for increased water and energy efficiency rather than NBSs.

The Municipal Operational Plan (Bologna Municipality 2009b, 2015c, 2015d) and the related SEA (Bologna Municipality 2015e) make 9 references to UES in terms of new greening actions for the different development areas defined. Cultural and regulating services are mentioned 6 and 3 times respectively. The SEA considers the same environmental sectors analysed in the PSC for the 28 development areas included in the Municipal Operational Plan. Greening interventions are mentioned to maintain and improve permeability and to face urban heat islands.

The Sustainable Urban Mobility Plan (Bologna Municipality 2018a, SUMP) includes 3 regulating UES: microclimate regulation, air pollution and noise reduction. It also provides recommendations to increase green in cycling and walking areas: 10% of the built area in new streets and 15% of the whole area in regenerated streets – pocket wetlands and rain gardens.

The Action Plan for Sustainable Energy (Bologna Municipality 2012, PAES) includes two measures that are related to UES: (a) recommendations to improve energy savings: green walls and roofs are not explicitly mentioned, but could be taken into account, and (b) the reference to the GAIA-EU project initiative, mentioned also in the Climate Adaptation Plan, which aims at increasing the number of trees planted in the city through direct agreements with local businesses.

Under the category called 'others', the documents that mentioned UES just once have been included. The Municipal code for public and private green (Bologna Municipality 2009a) recommends making use of tree species with good capacity of absorbing air pollutants in the development of new green areas. The Nature for children guidelines (Bologna Municipality 2015a) is the only document that recognises the potential educational services of urban nature in relation to child development. The urban farming and orchards code (Bologna Municipality 2009c) highlights the positive impact on social life and recreational activities that urban farming can facilitate, with a specific focus on the elderly.

10.3.3 Type of Action Related to UES

With this section, the study attempted to categorise the actions related to UES presented in paragraph 10.3.1.

The definition of the categories hereby presented arose from a qualitative content analysis and from informal interviews with city officers.

As shown in Fig. 10.3, most of the UES actions (61%) are expressed in the form of recommendations, that is, developers should include green roof, should improve accessibility, etc. Actions defined as infrastructures (26%) include concrete projects that the city will develop, that is, GAIA Life project and green spaces in defined development areas. Under governance/regulation, definitions of binding parameters were included that embrace actions related to urban standards for new developments and actions related to innovative governance schemes such as public–private partnership in the management of green areas. One action includes public financing scheme and it is related to agriculture funds for adaptation to climate change. Last, there is one action on awareness-raising, which refers to the Green-Up campaign on climate change and adaptation.

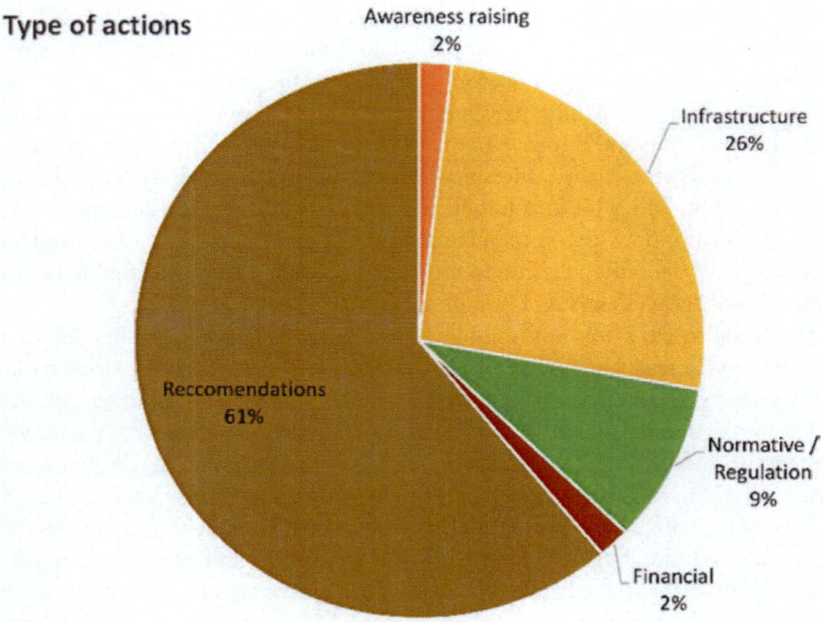

Fig. 10.3 Categories of actions including UES in the analysed document

10.4 Discussion and Conclusion

10.4.1 Current Situation

As mentioned in the results section, the Climate Adaptation Plan (Bologna Municipality 2015b) is the only document that makes explicit reference to the UES framework; however, UES have been referenced, in varying ways, in all the analysed documents. There could be several reasons for this inconsistence, including:

1. A large number of the documents were drafted at the beginning of 2000, when the ES framework was still relativity new and its potential application to the urban environment was unrealised; discussions related to the integration of UES into urban planning documents started to populate the literature at the start of the last decade.
2. Urban planning in Italy has a strong tradition of 'blueprint plans' that still makes the innovation proposed by the normative changes of the last 20 years difficult to be acknowledged in urban practice. The lack of interdisciplinary collaboration among different sectors, including planning, environmental and health departments, is one of the main obstacles to the innovation of urban planning tools.
3. The use of different periphrases and terms could sometime lead to misinterpretation of the concepts mentioned.

Nevertheless, the analysed documents contain a total of 58 references to UES framework. Those generally include regulating and recreational services with few references to provisioning and supporting services.

10.4.2 Areas for Development

Even though the analysed policy documents cited several UES and their potential positive impact on the city, the lack of an outstanding and overarching framework that would coordinate those requests is clearly a weak point of the overall urban strategy. Indeed, the improvement of UES should embrace a coordinated approach that could be boosted by the development of a dedicated green infrastructure strategy or through an overarching integration of the concept into existing local policies. Several actions and projects to boost UES are raising in the city, but the implementation of several single and not coordinated actions will hardly have a strong impact on the overall city quality.

Also, Bologna is currently lacking a comprehensive assessment of UES supply and demand. An UES assessment would support the city in defining overall targets and objectives, monitoring programme and ecosystem services-based decision support systems. Indeed, current targets and objectives, mentioned for example in the Climate Adaptation Plan (Bologna Municipality 2015b) and in the Action Plan for sustainable energy (Bologna Municipality 2012), are not fully integrated into urban planning documents. This lack of integration is then reflected in potentially effective but fragmented actions to be implemented and weak monitoring plans.

In this direction, the SEA could play a crucial role. Indeed, the analysis of each environmental component – that is, water, soil, electromagnetism, energy, mobility, air, noise and habitats for species – foresees a dedicated section to guarantee its sustainability. In this section, UES and NBSs are hardly mentioned while they could represent a great opportunity in several sectors, that is, water and air purification, noise pollution reduction, soil purification, etc. Indeed, after having identified the main challenges and opportunities in the urban plans, the SEA could give guidelines on the definition of NBSs to boost UES in the city.

10.4.3 Conclusion

Even though the overall framework of UES is not explicitly mentioned in Bologna's urban policies, except for the Climate Adaptation Plan (Bologna Municipality 2015b), regulating and cultural services are referenced in all the documents analysed. However, an overall framework with clear targets and monitoring programme to increase the quality of life and improve the urban environment is currently missing. Therefore, the UES framework represents a huge opportunity to plan a sustainable, resilient and healthy city.

The concept of UES is still abstract to most planners and practitioners in Italy, but it represents a strong framework for evaluation and assessment of plans and projects, providing decision makers with useful evidences regarding the environmental, social and economic benefit of such solutions. Therefore, this approach could push forward the implementation and the integration of NBSs and green infrastructure within spatial and urban planning.

Bologna is quite a compact city and faces challenges in introducing and building new green areas. An ES-based planning approach using NBSs represents a huge opportunity for Bologna and many other compact cities in Europe considering that NBSs can bring nature back into cities via a wide range of different solutions – that is, green roofs, green walls, green shelters, etc., and not just urban parks. It is therefore crucial to work on green urban regeneration, through greening of existing buildings, urban voids, and demolition and construction opportunities. In this sense, it is relevant to mention the new 24/2017 Emilia-Romagna urban planning law (Emilia Romagna Region 2017) that focuses on reducing land take and regeneration processes. The law demands cities to develop new City Master Plans (Piano Urbanistico Generale) before 2022, representing a huge occasion to integrate UES and NBSs into new strategies, plans and visions. Also, it could represent an opportunity to map UES supply and demand in the city area, starting from there to build up new priorities and targets.

Acknowledgements This work has been partially conducted within the implementation of the Urban Agenda on SUL and NBS Action Plan. The authors would like to thank the collaboration of several officers from the City of Bologna, in particular Dr. Eng, Giovanni Fini, for the great support and strong suggestions in the development of this work.

References

Baró F et al (2016) Mapping ecosystem service capacity, flow and demand for landscape and urban planning: a case study in the Barcelona metropolitan region, Land Use Policy. Pergamon 57:405–417. https://doi.org/10.1016/J.LANDUSEPOL.2016.06.006

Bolund P, Hunhammar S (1999) Ecosystem services in urban areas. Ecol Econ 29(2):293–301. https://doi.org/10.1016/S0921-8009(99)00013-0

Bologna Municipality (2007a) Piano Strutturale Comunale -PSC- Relazione ilustrativa

Bologna Municipality (2007b) Piano Strutturale Comunale -PSC- Valutazione di Sostenibilità ambientale e territoriale. Available at: http://www.comune.bologna.it/psc/articoli/1797

Bologna Municipality (2009a) Linee Guida per la progettazione delle aree verdi pubbliche. Available at http://www.comune.bologna.it/media/files/regolamento_comunale_del_verde_pubblico_e_privato__linee_guida.pdf

Bologna Municipality (2009b) Piano Operativo Comunale -POC – Presentazione e Norme. Available at: http://www.comune.bologna.it/psc/articoli/1797

Bologna Municipality (2009c) Regolamento per la conduzione e la gestione dei terreni adibiti ad aree ortive. Available at http://www.comune.bologna.it/media/files/regolamento_per_la_conduzione_e_la_gestione_dei_terreni_adibiti_ad_aree_ortive_1.pdf

Bologna Municipality (2012) Piano d'azione per l'energia sostenibile. Available at http://www.comune.bologna.it/paes

Bologna Municipality (2014a) Regolamento sulla collaborazione tra cittadini e amministrazione per la cura e la rigenerazione dei beni comuni urbani. Available at http://www.comune.bologna.it/sites/default/files/documenti/REGOLAMENTO%20BENI%20COMUNI.pdf

Bologna Municipality (2014b) Regolamento Comunale del verde Pubblico e Privato. Available at http://www.comune.bologna.it/media/files/regolamento_comunale_del_verde_testo_consolidato_280416.pdf

Bologna Municipality (2015a) Per restituire la natura ai bambini. Linee guida per la realizzazione di piccoli interventi naturali nei giardini dei nidi e delle scuole di infanzia di Bologna. Available at http://www.comune.bologna.it/sites/default/files/documenti/LINEE%20GUIDA%20GIARDINI%20.pdf

Bologna Municipality (2015b) Piano di adattamento città di Bologna. Available at http://www.blueap.eu/site/wp-content/uploads/2015/06/PianoBlueApfinale03062015.pdf

Bologna Municipality (2015c) POC- Rigenerazione di patrimoni Pubblici. Available at http://dru.iperbole.bologna.it/pianificazione-attuativa/poc-%E2%80%93-rigenerazione-di-patrimoni-pubblici

Bologna Municipality (2015d) POC- Variante programma per la qualificazione urbana diffusa. Available at http://dru.iperbole.bologna.it/pianificazione-attuativa/poc-%E2%80%93-programma-la-qualificazione-urbana-diffusa

Bologna Municipality (2015e) POC- Variante programma per la qualificazione urbana diffusa – Valutazione di sostenibilità ambientale e territoriale

Bologna Municipality (2018a) Piano Urbano della mobilità sostenibile. Available at: http://pums-bologna.it/

Bologna Municipality (2018b) Regolamento Urbanistico Edilizio. Available at: http://sit.comune.bologna.it/alfresco/d/d/workspace/SpacesStore/6da15be7-16dd-4f40-af43-056d24a965ab/Rue_vigore.pdf

Cortinovis C, Geneletti D (2018) Ecosystem services in urban plans: what is there, and what is still needed for better decisions, Land Use Policy. Elsevier Ltd 70:298–312. https://doi.org/10.1016/j.landusepol.2017.10.017

Costanza R, d'Arge R, de Groot R, Farber S, Grasso M, Hannon B, Limburg K, Naeem S, O'Neill RV, Paruelo J, Raskin RG, Sutton P, van den Belt M (1997) The value of the world's ecosystem services and natural capital. Nature 387(6630):253-260

Emilia Romagna Region (2017) Legge Regionale 21 Dicembre 2017 n.24

Frank B (2017) Urban systems: a socio-ecological system perspective. Sociol Int J 1(1). https://doi.org/10.15406/sij.2017.01.00001

Hansen R et al (2015) The uptake of the ecosystem services concept in planning discourses of European and American cities, Ecosystem Services. Elsevier 12:228–246. https://doi.org/10.1016/J.ECOSER.2014.11.013

Kabisch N (2015) Ecosystem service implementation and governance challenges in urban green space planning-the case of Berlin, Germany, Land Use Policy. Elsevier Ltd 42:557–567. https://doi.org/10.1016/j.landusepol.2014.09.005

Kaczorowska A et al (2016) Ecosystem services in urban land use planning: integration challenges in complex urban settings—case of Stockholm, Ecosystem Services. Elsevier B.V. 22:204–212. https://doi.org/10.1016/j.ecoser.2015.04.006

Mascarenhas A et al (2015) Ecosystem services in spatial planning and strategic environmental assessment-a European and Portuguese profile, Land Use Policy. Elsevier Ltd 48:158–169. https://doi.org/10.1016/j.landusepol.2015.05.012

Millennium Ecosystem Assessment (2005) Ecosystems and human well-being: biodiversity synthesis. World Resources Institute (WRI), Washington, DC

Nations, U et al (2018) World urbanization prospects the 2018 revision

Wilkinson C et al (2013) Strategic spatial planning and the ecosystem services concept – an historical exploration, ecology and society. Resilience Alliance 18(1). https://doi.org/10.5751/ES-05368-180137

Woodruff SC, BenDor TK (2016) Ecosystem services in urban planning: comparative paradigms and guidelines for high quality plans, landscape and urban planning. Elsevier 152:90–100. https://doi.org/10.1016/j.landurbplan.2016.04.003

Zanon B, Verones S (2013) Climate change, urban energy and planning practices: Italian experiences of innovation in land management tools, land use policy. Elsevier Ltd 32:343–355. https://doi.org/10.1016/j.landusepol.2012.11.009

Chapter 11
The New Urban Plan of Rescaldina Municipality. An Experience for Improving Ecosystem Services Provision

Silvia Ronchi, Andrea Arcidiacono, and Laura Pogliani

Abstract This contribution presents the results of an urban planning process experience conducted in the Municipality of Rescaldina (located in the northern Milan metropolitan area, Lombardy region, north-west of Italy).

The new local Urban plan (including the Strategic Environmental Assessment – SEA) was developed based on ecosystem services (ES) addressing urban transformations and future development strategies towards the improvement of their performance as well as the promotion of human health and well-being. The integration of an ES-based approach in Planning involved the deployment of a local green infrastructure (GI) as the backbone for the design of urban and suburban public and private spaces, using nature-based solutions. The ES-based GI is a core strategy of both the SEA and the Urban plan, the assumption into both strategic and prescriptive frameworks of the Urban plan ensures its operability.

Keywords Urban planning · Urban ecosystem services · Regulating services · Open spaces design · Urban green infrastructures · Land-use regulation · Decision-making process · Performance-based planning · Human well-being

---

Text and graphics of this chapter are based on Ronchi S, Arcidiacono A, Pogliani L (2020) Integrating green infrastructure into spatial planning regulations to improve the performance of urban ecosystems. Insights from an Italian case study. Sustainable Cities and Society 53:1–12. https://doi.org/10.1016/j.scs.2019.101907.

S. Ronchi (✉) · A. Arcidiacono · L. Pogliani
Department of Architecture and Urban Studies, Politecnico di Milano, Milano, Italy
e-mail: silvia.ronchi@polimi.it

A. Arcidiacono, S. Ronchi (eds.), *Ecosystem Services and Green Infrastructure*, Cities and Nature, https://doi.org/10.1007/978-3-030-54345-7_11

11.1 A New Urban Planning Paradigm

In 2014, the newly elected local administration of Rescaldina – a municipality with more than 14,000 inhabitants located north of Milan – decided to start a radical revision of the local Urban plan although the previous one was recently approved (in 2012).[1]

This decision stems from the divergence between the new administration's electoral programme and the development policies set out in the 2012 Urban plan, which included a huge amount of new transformation areas, for commercial, industrial and residential settlements, with a significant alteration in the provision of ecosystem services (ES). This alteration strongly depends on soil sealing and land take dynamics, as the 2012 Urban plan forecasts the transformation of more than 210,000 square metres of agricultural and natural areas in artificial surfaces.

In addition to this prevision, another significant Urban transformation area was planned in the municipality of Cerro Maggiore, bordering Rescaldina to the south, involving 300,000 square metres of agricultural land for a new large-scale commercial area. This transformation would have had heavy implications in terms of habitat loss, ecological fragmentation, landscape and ecosystem degradation challenging to compensate.

Due to these critical issues, the new Rescaldina's administration initiated a general revision to 2012 Urban plan pursuing its winning electoral programme objectives, that is, defining a participatory, sustainable and resilient Urban plan, by ecologically and environmentally enhancing the natural and agricultural system; implementing a widespread quality of public spaces and urban greenery; strengthening of soil permeability by limiting soil sealing and land take process and achieving hydraulic invariance; improving the soft mobility with new local and supralocal paths, also developing a "green mobility" plan; redeveloping degraded areas and regenerating brownfields; and supporting local businesses, particularly threatened by a recent large-scale shopping mall located in the southern part of the municipality (Arcidiacono et al. 2018a).

Aiming to experiment different planning approach, based on land take limitation, combined with an Urban plan able to respond to the new environmental, ecological and social needs, the Municipal Administration requested a scientific and technical support to the Department of Architecture and Urban Studies (DAStU) of Politecnico di Milano finalised to prepare studies and research, fact-finding and interpretative surveys on the quality of open spaces and urbanisation dynamics (through an ES mapping) and support the municipality planning department to define strategies and structurally design the Plan (and related Strategic Environmental Assessment – SEA).

A further operational request made by the local government was to have a "unique, comprehensive strategy for land-use regulation which could address

[1] In the Lombardy Region, the Urban plan is called: *Piano di governo del territorio,* in English:Territorial development plan.

multiple issues, including public space design, natural landscape quality, regulation of land use in peri-urban areas, increased urban resilience as part of climate change adaptation" (Ronchi et al. 2020). This request has significantly guided and addressed the ecosystem services integration into the Urban plan.

The new Urban plan of Rescaldina was approved in March 2019, and it is still in force.

11.2 Understanding the Planning System to Integrate Ecosystem Services in Urban Plans

The proposed methodology is innovative for the adoption of an ES-based green infrastructure (GI), as a design tool which provides a multiplicity of ecosystem services based on their functions (Arcidiacono et al. 2018b), for planning purposes.

GI can be defined as a "network of natural and semi-natural areas with other environmental features that is supposed to deliver ecosystem services" (European Commission 2013) and as "a design vision that translates [a] planning strategy into physical reality while heeding the ecological and cultural characteristics of a particular locale – whether a region or an individual building" (Rouse and Bunster-Ossa 2013). As stated by Benedict and McMahon (2001), and by Hansen and Pauleit (2014), five principles guide the GI: "1) integration: considering the grey–green combination of GI; 2) multifunctionality: GI includes the ecological, social and economic/abiotic, biotic and cultural functions of green spaces; 3) connectivity between green spaces; 4) multiscale approach taking in all parcels, from the individual to the community, regional and state scale; 5) multi-object approach including diverse types of (urban) green and blue space."

GI partially takes up the ecological network concept and biodiversity targets but it emphasises the multifunctionality of the ecosystems and uses nature-based solutions (NBS), that is, "living solutions inspired and supported by the use of natural processes and structures [which] are designed to address various environmental challenges in an efficient and adaptable manner, while simultaneously providing economic, social, and environmental benefits" (European Commission 2015; Maes and Jacobs 2017).

GI represents the framework for the contemporary city urban design as a supporting structure (i.e. a backbone) useful to evaluate and verify the plan's urban transformation choices, addressing them to environmentally sustainable solutions for enhancing well-being. The adoption of an ecosystem approach to a local GI requires to overcome the traditional urban planning paradigm. As argued by Ronchi et al. (2020), traditional land-use planning was "based exclusively on land use (whether residential, manufacturing or commercial), these models do not take into account the suitability of the land to host a specific function or its consistency within a wider territorial context." Historically, the design of a city's green spaces was guaranteed by quantitative standards which set a mandatory minimum share of public services. The Italian Inter-Ministerial Decree n. 1444/1968 sets the mandatory amount of 18 sqm per inhabitant in the sizing of the areas as "Planning standards." Such standards are to be considered the minimum provision of public facilities and public open spaces at the neighbourhood and local scale, subdivided into 4.50 sqm for education facilities, 2 sqm for facilities of common interest, 9 sqm for urban green spaces and 2.50 sqm for parking areas (Italian Government 1968).

The adoption of an ecosystem approach requires a different planning model based on the qualitative performance of multiple benefits in terms of regulation, support, provision and cultural services.

The innovation aspect of this research lies in the GI operability, which is guaranteed by the integration of the ecosystem assessments and GI design into the Urban plan with different forms of applications. This aspect is essential to transform the academic and scientific studies into a tool which is useful to the policymaker, enabling choices aimed at increasing the supply of ES to ensure a better quality of life for citizens, and support ordinary urban planning activities.

Understanding how to implement ES for planning purpose, it is important to know the planning system, the municipal urban planning tools and how they work and what effects they could produce on ES provision. Otherwise, the risk is that ES integration is a mere declaration of intent, often a recommendation (Haase et al. 2014) without a practical impact on the supply of ES or their management (Hansen and Pauleit 2014; Geneletti et al. 2017; Ronchi et al. 2020).

Considering this premise, the Lombardy urban planning system is explained in the following text, that aims to understand the process of integration of the research into the Plan and in all its components (strategic, operational and prescriptive).

The Lombardy Regional law on *Governo del territorio* (Territorial Government) n. 12 of 2005 sets the structure and process that local municipalities follow on preparing the Urban plan named Territorial Development Plan (TDP). The TDP is composed of three documents:

1. *Documento di Piano* (Planning document – DdP) contains a general framework of strategies, analysis, objectives and guidelines for the territorial, social and economic development of a municipality. It is valid for 5 years (this time is strictly related to the local Mayor's electoral period), can be modified at any time and contains indications that have no effects on the land-use rights. The DdP sets the new Urban Transformation (UT) "as areas, generally natural or agricultural, converted to host human activities (i.e. residential, commercial and tertiary)" (Ronchi et al. 2020).
2. *Piano delle Regole* (Regulative Plan – PdR) regulates the existent urbanised city. It has no temporal limit, can be modified at any time with direct effects on the land-use rights and property.
3. *Piano dei Servizi* (Services Plan – PdS) tackles with the issue of local services at municipal level considering not only the quantitative supply of areas and facilities but also the quality of services (in terms of performance, accessibility, efficiency and financial feasibility) in relation with the demand, the composition of the population and the different types of needs expressed, aiming to enhance the

quality and urban liveability. The PdS is drawn up to achieve the requirements of habitability and urban quality through the concept of public service. It has no temporal limit, can be modified at any time and deals specifically with the planning and the design of public services and facilities (such as social housing and green areas).

Considering this triple division (DdP, PdS and PdR) of the Urban plan in the Lombardy region, the local GI of Rescaldina was included in each one. The underlying goals are to improve the natural capital and human well-being through the conditioning and regulation of the existent public and private city and to address the transformation areas towards sustainable development. Specifically, the DdP fixes precise design strategies, also based on NBS, for the development of the forecasted UT following the ES-based GI strategy.

As stated by Ronchi et al. (2020), NBS are greening design actions that can contribute to developing GI in urban areas while GI is an application-oriented tool for integrating ES concept into land-use planning. For the TDP of Rescaldina, a specific catalogue of NBS was created according to GI landscape type and to the three management strategies (see Subchapter 11.3) including, for example, the creation of shrubs, woodlands, wetlands, green roofs, rain gardens, rural pathways and tree lines.

The UT criteria and guidelines include priority interventions for the public city and urban and environmental equipment (in accordance with the SEA and PdS) as dedication areas, as grant of private land for public use; riparian buffer zones; public spaces with permeable surfaces and facilities; private space for urban orchards; urban forestry; new pathways; green parking lots with permeable or semi-permeable paving.

The same approach was used for the projects concerning the so-called "public city," the one disciplined by PdS, identifying specific design criteria to improve the quality of public spaces based on ES performance including them in the overall strategy of the GI. Moreover, in the PdS, a sample of design schemes was provided to suggest how the NBS can be implemented in the public city.

Lastly, the PdR sets the rules for the built-up city giving precise prescriptions for each land-use classes (residential, industrial, mixed-use, tertiary, etc.) according to the ES-based GI. As an example, for the low-density residential areas, the PdR prescribes the urban forestry of native species in private green areas, the enhancement of the permeable open spaces, or the prohibition of cutting down of trees without a valid reason (Fig. 11.1).

The inclusion of GI strategy in the three documents of TDP ensures the complete transposition of the GI project into all the steps of the planning process, from the future vision of the territory to the land-use regulations, in the public and private domain.

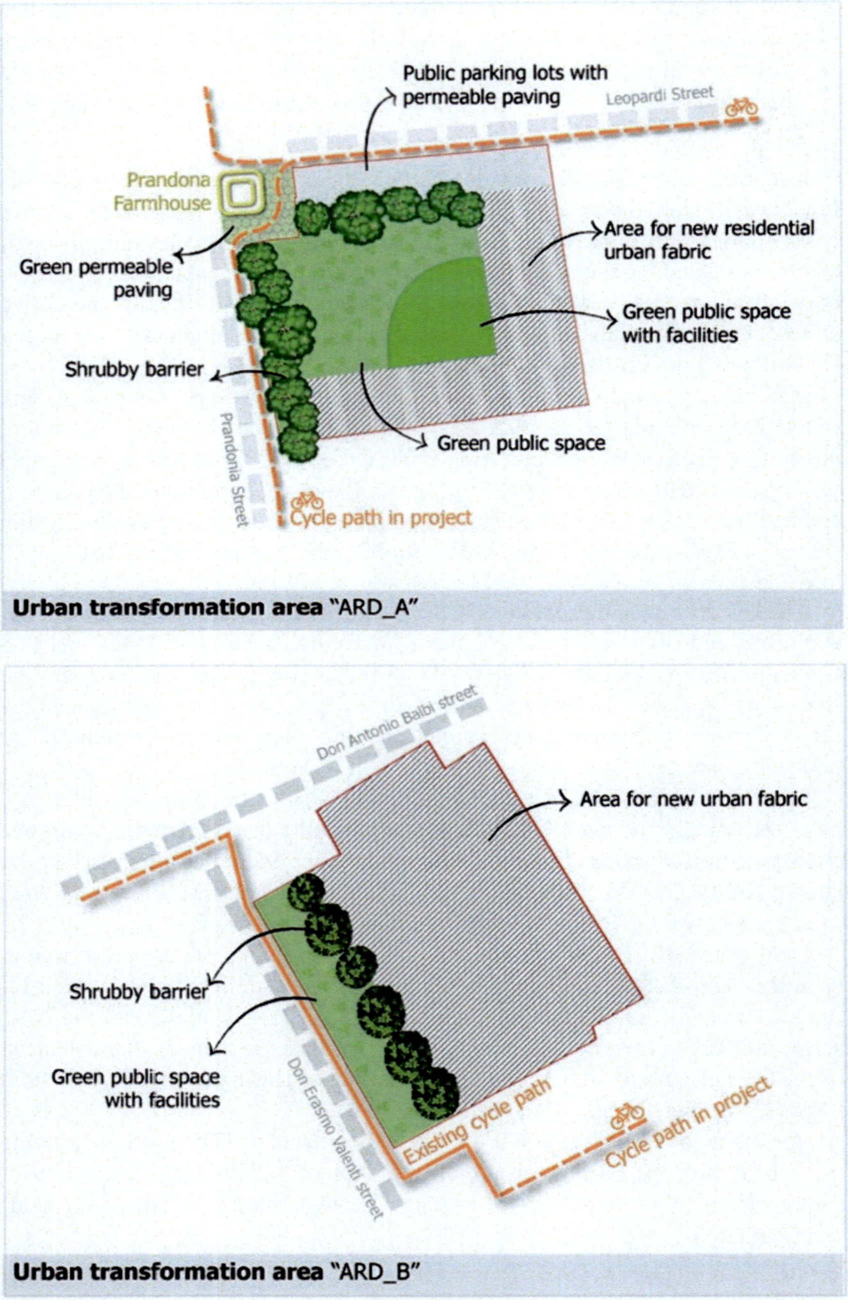

Fig. 11.1 Sample schemes of two UT area. Nature-based solutions are used for the design of green public spaces addressing urban built-up development (Based on Ronchi et al. 2020)

11.3 Guiding the Planning Process Through Ecosystem Services Mapping and Assessment

An important support for configuring the GI design and for the definition of urban regeneration strategies comes from the evaluation and mapping of ecosystem functions and related services (Naidoo et al. 2008; Burkhard et al. 2012, 2013) as "benefits that humans obtain from ecosystem functions" (de Groot et al. 2002; Millennium Ecosystem Assessment 2005) or "as direct and indirect contributions from Ecosystems to Human Well-being" (ten Brink et al. 2009). At the least, ES are the set of processes and conditions that make possible the survival of human life in Natural Ecosystems (Ronchi 2018).

The methodology identified for the construction of the Rescaldina GI was developed within the Strategic Environmental Assessment (SEA) process starting from the combined mapping and analysis of ecosystem functions as recognised by the Common International Classification of Ecosystem Services (CICES) (Haines-Young et al. 2018).

The first 5 layers – on regulating and supporting services – were elaborated using InVEST (Integrated Valuation of Ecosystem Services and Tradeoffs), an open-source software, developed during the Natural Capital Project and jointly developed by Stanford University, the University of Minnesota, the Nature Conservancy and the World Wildlife Fund. The software is specifically dedicated for regional and urban planning considering both economic and ecological accounting (Tallis et al. 2011; Arcidiacono et al. 2016; Ronchi and Arcidiacono 2018). The InVEST outputs were subsequently processed using the ESRI ArcGIS platform to perform a weighted overlay analysis.

The ES modelled are the following:

1. Habitat quality, measured in terms of overall ecological quality based on proximity of the habitat to artificial land uses and the degree of disturbance caused by them (Tallis et al. 2011; Salata et al. 2017)
2. Carbon sequestration as the quantity of carbon stocked in 4 primary pools (above-ground biomass, below-ground biomass, soil, dead organic matter) (Tallis et al. 2011; Arcidiacono et al. 2015)
3. Water yield as annual water yield from a catchment area with the intended end use of reservoir hydropower production (Tallis et al. 2011; Redhead et al. 2016)
4. Sediment retention, that is, "the capacity of a land parcel to retain sediment by using information on geomorphology, climate, vegetative coverage and management practices" (Tallis et al. 2011);
5. Soil erosion, based on the Universal Soil Loss Equation (USLE), an empirical equation used to predict average annual erosion (Wischmeier and Smith 1978)

The cultural service was evaluated as the Cultural heritage distribution, selected according to Italian Legislative Decree no. 42 of 2004 concerning Cultural Heritage and Landscape, considered as an "important aspect of cultural and amenity services as a whole, implying the non-material benefits people obtain from ecosystems through spiritual enrichment; cognitive, emotional and social development; reflection; recreation; and aesthetic experiences" (Millennium Ecosystem Assessment 2005). It has been estimated using a kernel density function in order to obtain a spatial concentration of heritage sites (Fig. 11.2).

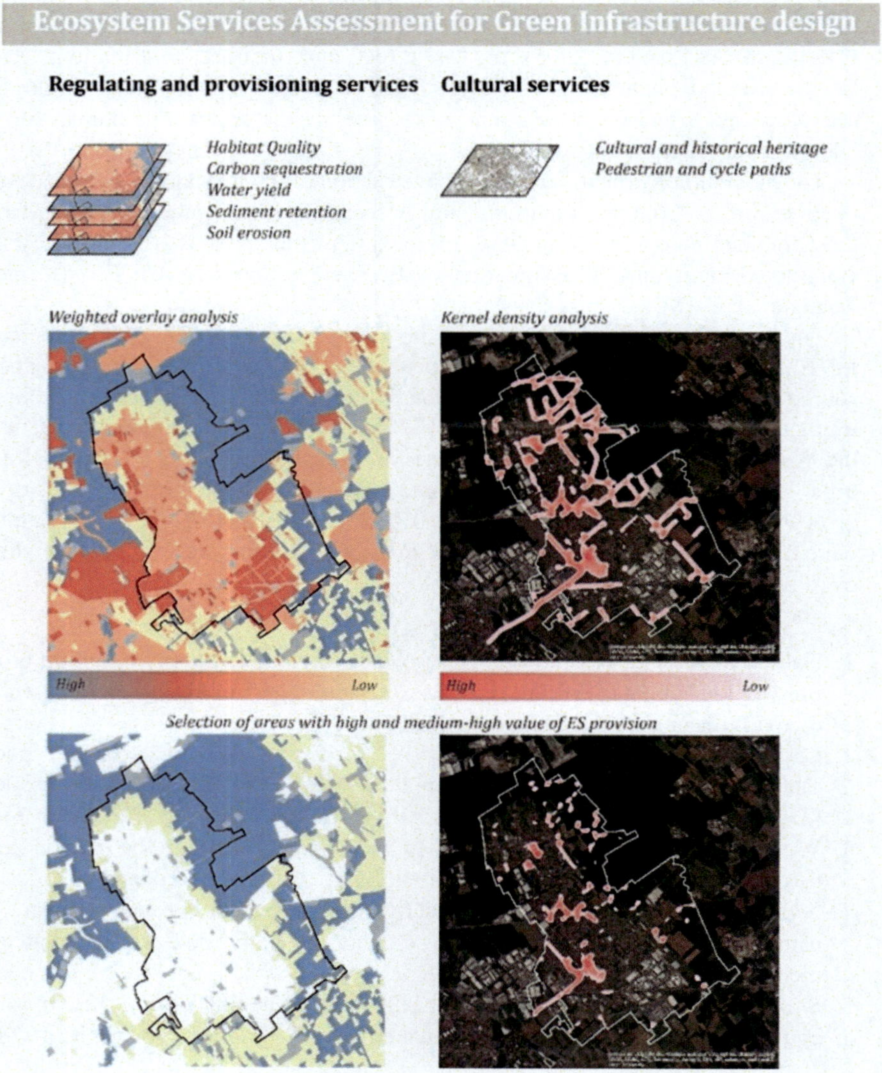

Fig. 11.2 Ecosystem services assessment for green infrastructure design (divided by ES type: regulating services, provisioning services and cultural services) (World imagery sources: Esri, DigitalGlobe, Earthstar Geographics, CNES/Airbus DS, GeoEye, USDA FSA, USGS, Getmapping, Aerogrid, IGN, IGP, and the GIS User Community) (Based on Ronchi et al. 2020)

The combination of the analyses allows to define the most statistically significant areas and to select the ones to be included in the GI strategy. This selection involves areas with high ES supply values, which must be preserved and protected, and also degraded or abandoned areas that need to be regenerated and restored for improving their ecosystem performance in line with what is defined by target 2 of the European Biodiversity Strategy (European Commission 2011).

Assuming the methodology defined in the proposed Lombardy Regional Landscape Plan (mentioned in Chap. 5) (Arcidiacono et al. 2016; Salata et al. 2016) and as stated by Ronchi et al. (2020), the Rescaldina GI is divided into three types of landscape:

1. "Natural landscapes, including forests and semi-natural areas, deemed important for biodiversity and conservation reasons (in Rescaldina these areas mainly coincide with the wooded area named *Bosco del Rugareto*)
2. Anthropic landscapes, featuring historical and cultural heritage sites (mainly derived by the cultural ES assessment corresponding to the historical city centres)
3. Rural landscapes, featuring elements of traditional rural landscapes and consisting of "mosaics" of small-scale arable fields, traditional paddy fields and linear rural elements (Ciaian and Paloma 2011) (these mainly coincide with peri-urban areas characterised by a medium-to-high presence of regulating and provisioning services)"

For each landscape category (natural, anthropic or rural), three different levels of ES provision have been identified, based on the above analyses. They give rise to three different types of actions/strategies (maintenance, valorisation and regeneration). The maintenance strategies have been designed for areas with a high ecosystem value which need to be protected avoiding depletion and degradation that could compromise their quality. For the areas with a medium ES value, the actions are oriented towards the improvement of their performance, or in the worst case, the preservation of the current ecosystem quality. Regeneration and restoration strategies involve degraded areas such as quarries, brownfields, construction sites and landfills, for which it is necessary to reverse the current ES state and trends (Fig. 11.3).

The GI represents a strategy for Rescaldina municipality able to address a multiplicity of issues, including public space design, natural landscape quality, regulation of land use in peri-urban areas, increase urban resilience in climate change adaptation, suggesting actions and solutions based on territorial vocations and performances.

11.4 Conclusions

The adoption of an ecosystem approach for GI deployment guided the overall planning experience of Rescaldina municipality.

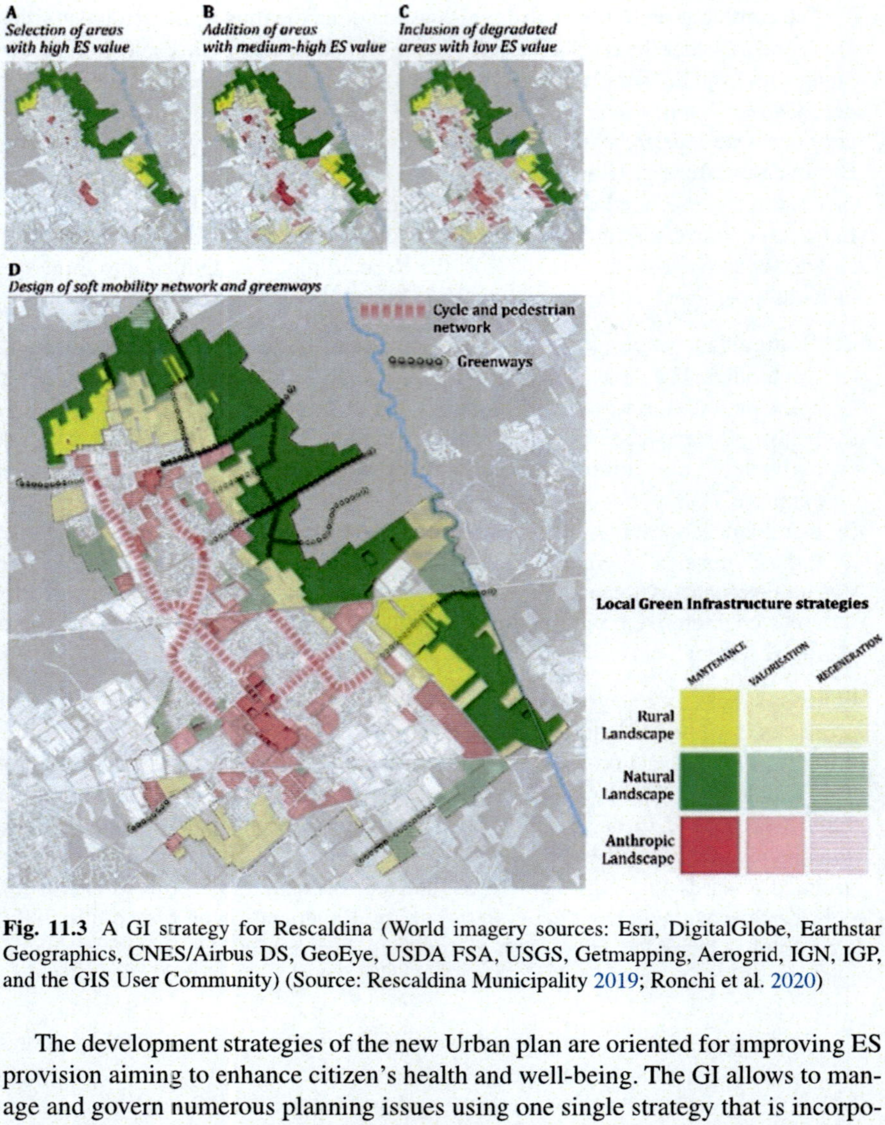

Fig. 11.3 A GI strategy for Rescaldina (World imagery sources: Esri, DigitalGlobe, Earthstar Geographics, CNES/Airbus DS, GeoEye, USDA FSA, USGS, Getmapping, Aerogrid, IGN, IGP, and the GIS User Community) (Source: Rescaldina Municipality 2019; Ronchi et al. 2020)

The development strategies of the new Urban plan are oriented for improving ES provision aiming to enhance citizen's health and well-being. The GI allows to manage and govern numerous planning issues using one single strategy that is incorporated in the Urban plan and all its components and articulations. The operability of GI is guaranteed by its integration in the regulative tool with some mandatory actions for orienting new UT areas and the existing urbanised city promoting the adoption of NBS. GI advocates ES in Spatial planning using NBS to improve the performance of the urban design.

This research experience shows how ES could be integrated into the planning process overcoming the ordinary approach towards a performance-based one highlighting solutions and opportunities and aiming to bridge the science–policy interface.

Acknowledgements We are grateful to the research group that worked on this project (Athos Brenna, Viviana di Martino, Federico Jardini, Silvia Restelli and Stefano Salata) and to the local government officials (especially Marianna Laino, deputy councillor for Urban Planning) who generously gave their valuable time to contribute to the research.

References

Arcidiacono A, Ronchi S, Salata S (2015) Ecosystem services assessment using InVEST as a tool to support decision making process: critical issues and opportunities. In: Computational science and its applications -ICCSA 2015. Springer International Publishing, Cham, pp 35–49

Arcidiacono A, Ronchi S, Salata S (2016) Managing multiple ecosystem services for landscape Conservation: a green infrastructure in Lombardy Region. In: Procedia Engineering, p 2297–2303

Arcidiacono A, Pogliani L, Ronchi S (2018a) Contenere il consumo di suolo attraverso il progetto urbanistico. Il disegno della rete verde quale struttura strategica nel PGT del comune di Rescaldina (Mi). In: Arcidiacono A, Di Simine D, Ronchi S, Salata S (eds) Consumo di suolo, servizi ecosistemici e green infrastructures: Caratteri territoriali, approcci disciplinari e progetti innovativi. Rapporto 2018 CRCS. INU Edizioni, Roma, pp 138–150

Arcidiacono A, Ronchi S, Salata S (2018b) Un approccio ecosistemico al progetto delle infrastrutture verdi nella pianificazione urbanistica. Sperimentazioni in Lombardia. In: Urbanistica. INU Edizioni, Roma, pp 102–113

Benedict MA, McMahon ET (2001) Green infrastructure: smart conservation for the 21st century. Renew Resour J 20:12–17

Burkhard B, Kroll F, Nedkov S, Müller F (2012) Mapping ecosystem service supply, demand and budgets. Ecol Indic 21:17–29. https://doi.org/10.1016/j.ecolind.2011.06.019

Burkhard B, Crossman N, Nedkov S et al (2013) Mapping and modelling ecosystem services for science, policy and practice. Ecosyst Serv 4:1–3. https://doi.org/10.1016/j.ecoser.2013.04.005

Ciaian P, Paloma SGY (2011) The value of EU agricultural landscape. Publications Office of the European Union, Luxembourg

de Groot R, Wilson MA, Boumans RMJ (2002) A typology for the classification, description and valuation of ecosystem functions, goods and services. Ecol Econ 41:393–408. https://doi.org/10.1016/S0921-8009(02)00089-7

European Commission (2011) The EU biodiversity strategy to 2020. Publications Office of the European Union, Luxembourg

European Commission (2013) Building a green infrastructure for Europe. Publications Office of the European Union, Luxembourg

European Commission (2015) Towards an EU research and innovation policy agenda for nature-based solutions & re-naturing cities. Publications Office of the European Union, Luxembourg

Geneletti D, La Rosa D, Spyra M, Cortinovis C (2017) A review of approaches and challenges for sustainable planning in urban peripheries. Landsc Urban Plan 165:231–243. https://doi.org/10.1016/j.landurbplan.2017.01.013

Haase D, Larondelle N, Andersson E et al (2014) A quantitative review of urban ecosystem service assessments: concepts, models, and implementation. Ambio 43:413–433. https://doi.org/10.1007/s13280-014-0504-0

Haines-Young R, Potschin-Young M, Czúcz B (2018) Report on the use of CICES to identify and characterise the biophysical, social and monetary dimensions of ES assessments

Hansen R, Pauleit S (2014) From multifunctionality to multiple ecosystem services? a conceptual framework for multifunctionality in green infrastructure planning for urban areas. Ambio 43:516–529. https://doi.org/10.1007/s13280-014-0510-2

Italian Government (1968) DI n. 1444 Limiti inderogabili di densità edilizia, di altezza, di distanza fra i fabbricati e rapporti massimi tra gli spazi destinati agli insediamenti residenziali e produttivi e spazi pubblici o riservati alle attività collettive, al verde pubblico o a parcheggi, da osservare ai fini della formazione dei nuovi strumenti urbanistici o della revisione di quelli esistenti, ai sensi dell'art. 17 della legge n. 765 del 1967

Maes J, Jacobs S (2017) Nature-based solutions for Europe's sustainable development. Conserv Lett 10:121–124. https://doi.org/10.1111/conl.12216

Millennium Ecosystem Assessment (2005) Ecosystems and human Well-being: synthesis. Island Press, Washington, DC

Naidoo R, Balmford A, Costanza R et al (2008) Global mapping of ecosystem services and conservation priorities. Proc Natl Acad Sci U S A 105:9495–9500. https://doi.org/10.1073/pnas.0707823105

Redhead JW, Stratford C, Sharps K et al (2016) Empirical validation of the InVEST water yield ecosystem service model at a national scale. Sci Total Environ 569–570:1418–1426. https://doi.org/10.1016/j.scitotenv.2016.06.227

Rescaldina Municipality (2019) Valutazione ambientale strategica. Rapporto Ambientale. Available at: http://pgt.rescaldina.org

Ronchi S (2018) Ecosystem Services for Spatial Planning. Innovative approaches and challenges for practical applications. Green Energy Springer International Publishing AG, Part of Springer Nature 2018, Cham

Ronchi S, Arcidiacono A (2018) Adopting an ecosystem services-based approach for flood resilient strategies: the case of Rocinha Favela (Brazil). Sustain. https://doi.org/10.3390/su11010004

Ronchi S, Arcidiacono A, Pogliani L (2020) Integrating green infrastructure into spatial planning regulations to improve the performance of urban ecosystems. Insights from an Italian case study. Sustain Cities Soc 53:1–12. https://doi.org/10.1016/j.scs.2019.101907

Rouse DC, Bunster-Ossa IF (2013) Green infrastructure: a landscape approach. American Planning Association, Washington, DC

Salata S, Ronchi S, Ghirardelli F (2016) I servizi ecosistemici a supporto della pianificazione paesaggistica. Territorio 77:45–52

Salata S, Ronchi S, Arcidiacono A, Ghirardelli F (2017) Mapping habitat quality in the Lombardy Region, Italy, One Ecosyst 2: e11402

Tallis HT, Ricketts T, Guerry AD, et al (2011) InVEST 2.0 beta user's guide. Stanford

ten Brink P, Kettunen M., Vakrou A, Wittmer H (2009) The economics of ecosystems and biodiversity. TEEB for National and International Policy Makers

Wischmeier W, Smith D (1978) Predicting rainfall erosion losses. U.S. Department of Agriculture, Washington DC

Chapter 12
Identifying Ecosystem Service Hotspots to Support Urban Planning in Trento

Davide Geneletti and Chiara Cortinovis

Abstract This chapter presents the mapping of ecosystem service hotspots in the city of Trento, in the Italian Alps, and discusses the possible uses of the results to support the drawing of the new Urban Plan. Hotspots are defined as areas characterized by high levels of provision of multiple services. Particularly, the following ecosystem services were selected: biodiversity support, risk mitigation, mitigation of air pollution and noise from transport infrastructures, forest landscape value, food production, nature-based recreation, and microclimate regulation. Maps of the different ecosystem services were produced using a variety of assessment methods, and then overlaid to obtain a multiple-hotspot map. The results show that hotspots are found not only among forests and natural areas surrounding the city, but also in the intensely urbanized valley floor (e.g., within agricultural patches and green wedges). In terms of policy relevance, the ecosystem service hotspots are becoming part of the "structural elements" of the Urban Plan currently being drafted, along with more traditionally recognized elements, such as protected areas. The hotspots are intended to be preserved from urbanization and different actions are under consideration to improve the current network of green and blue spaces, thus increasing both connectivity and the provision of ecosystem services.

Keywords Ecosystem service hotspots · Urban planning · Green infrastructure · GIS · Air purification · Noise reduction · Cooling · Nature-based recreation · Habitat provision · Risk mitigation

---

D. Geneletti (✉)
Department of Civil, Environmental and Mechanical Engineering, University of Trento, Trento, Italy
e-mail: davide.geneletti@unitn.it

C. Cortinovis
Centre for Environmental and Climate Research, Lund University, Lund, Sweden

Department of Civil, Environmental and Mechanical Engineering, University of Trento, Trento, Italy

A. Arcidiacono, S. Ronchi (eds.), *Ecosystem Services and Green Infrastructure*, Cities and Nature, https://doi.org/10.1007/978-3-030-54345-7_12

12.1 Introduction

Trento is an alpine city of 120,000 inhabitants in northeastern Italy. The main settlement is located along the Adige River's valley floor, and hosts around 70% of the city's population. The remaining 30% lives in small villages spread on the close-by hillsides. The municipal area is large (around 156 km) and includes part of the mountain slopes surrounding the city, up to an elevation of 2180 m.

According to municipal data, urban and peri-urban green areas account for 210 ha, 157 of which are usable green spaces (13.4 m^2/inhabitant), while forests cover around 1/3 of the municipal territory. This determines a generally high proximity of urban areas to natural environments. More than 10 km^2 of the city administrative area is designated as natural protected area, including eight Natura 2000 sites and four local reserves. The municipal territory also includes the largest total agricultural area of all municipalities in the province. The main cultivations are vineyards and apple orchards, which occupy sectors of the hills and the few non-urbanized patches in the valley floor (Fig. 12.1).

In the last years, one of the main objectives of the administration has been to increase the amount of public green areas within the city, particularly in deprived neighborhoods, so as to improve equity in distribution and access. Recently, most of the efforts have focused on peri-urban areas, by creating new parks and launching activities aimed at improving the daily practices of green space management through citizens' involvement in co-designing and co-developing new functions and uses.

The drafting of the new Urban Plan for Trento, which was initiated in 2017 and it is still ongoing, provided the opportunity to revise, update, and coordinate strategies and initiatives regarding urban green infrastructure. As part of the process, an urban ecosystem service assessment was carried out, with the overall purpose of understanding how actions and instruments of the Urban Plan could enhance the provision of ecosystem services and related benefits in the city. Among the potential uses of ecosystem service knowledge to support the planning process, a key policy question emerged as follows: How can ecosystem service assessments improve the identification of the structural elements of the Urban Plan?

This question has been addressed by identifying ecosystem service "hotspots", that is, areas characterized by high levels of provision of multiple services (García-Nieto et al. 2013; Geneletti et al. 2018). This chapter presents the mapping of ecosystem service hotspots in the city of Trento and discusses the possible uses of the results to support the drawing of the new Urban Plan.

12.2 Selecting and Assessing Ecosystem Services

The following ecosystem services were selected: biodiversity support, risk mitigation, mitigation of air pollution and noise from transport infrastructures, forest landscape value, food production, nature-based recreation, and microclimate regulation

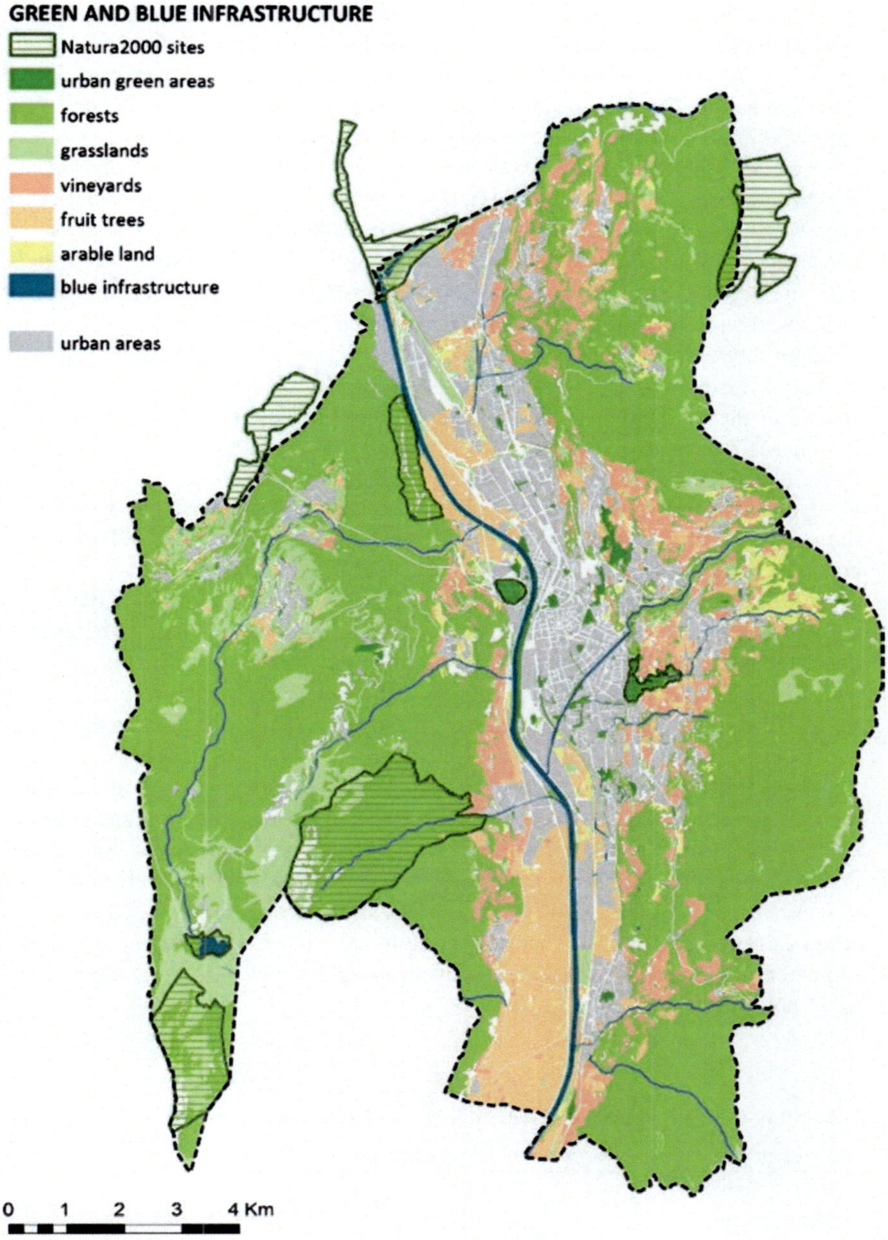

Fig. 12.1 The city of Trento and its green and blue infrastructure

(cooling). The ecosystem services were selected in close collaboration with key staff from city administration, who contributed to the definition of the policy questions, and provided feedback on the results. The selection is also consistent with the main issues that emerged from the strategic document that the city administration approved in 2018 to steer the urban planning process.

Given the objectives of the work and the need to effectively using and updating the results during planning exercises, it was decided to limit the complexity of the ecosystem service assessment as much as possible. In some cases, the mapping consisted in the selection of suitable proxies, either elements already identified in existing thematic maps or areas relevant for the supply of a specific ecosystem service based on the definition of indicators and thresholds. In other cases, the mapping required the reprocessing of existing data, also through the application of models.

The assessment of ecosystem services was carried out separately for two distinct sectors of the municipality: the wooded areas in the slopes surrounding the valley floor, and the peri-urban and valley-bottom green areas. The latter includes the agricultural areas surrounding the settlements on the slopes, the agricultural areas on the valley floor, some wooded fragments (or otherwise separated from the main areas of the slopes), and the urban green areas. Table 12.1 shows the ecosystem services considered for each of the two sectors.

All the analyses were carried out in a GIS environment using the QGIS v.2.18 and GRASS v.7.2 software. The results are maps that can be easily integrated into the databases of territorial data of the municipality, and used as inputs for further analysis. The format also allows their possible integration with webGIS applications. Following is a brief description of the approach adopted to assess the selected ecosystem services, summarized in Table 12.2.

Biodiversity support considers the role of ecosystems as a habitat for species (supporting ecosystem service). Biodiversity is at the origin of ecosystem services. Protecting biodiversity and preventing the exploitation of ecosystem services from degrading the areas and natural elements that provide them are therefore necessary to guarantee sustainable supply of ecosystem services over time. Specifically in the case of Trento, the analysis focused on the identification, in the municipal area, of protected areas and other habitats relevant to animal species, using existing data (see Table 12.2).

Table 12.1 Ecosystem services considered for each of the two sectors

Ecosystem service	Wooded slopes	Green areas in the valley floor
Habitat for biodiversity	X	X
Risk mitigation	X	X
Air purification and noise reduction	X	X
Landscape value	X	
Food production		X
Nature-based recreation	X	X
Cooling		X

Table 12.2 Overview of the approach followed to assess ecosystem services and identify hotspots

Ecosystem service	Indicator	Hotspot (values/thresholds)
Habitat for biodiversity	Potential richness of focal species (Pedrini et al. 2013)	Biodiversity hotspots (i.e., areas larger than 200 ha with high potential richness) medium, high, or very high potential richness
	Level of protection	Protected areas at various levels (Natura 2000 sites, local reserves, etc.)
Risk mitigation (rock falls, landslides, floods)	Protective function with respect to rock falls (Wolynski et al. 2016)	Forest patches characterized by potential protective function
	Level of landslide hazard	Forest patches in areas of high geological hazard
	Level of river flooding hazard	Permeable (non-urban) areas in flood-prone zones
Air purification and noise reduction	Proximity to high-traffic roads (and railways for noise) and residential settlements	Wooded areas within a maximum distance of 50 m from both residential settlements and main transport infrastructures
Landscape value (aesthetic appreciation and identity)	Class of landscape value based on forest types and location	High landscape value
Food production	Suitability for agriculture (a combined indicator considering economic value based on current crops and morphological suitability)	Suitability ≥ 6 (range: from 2 to 8)
Nature-based recreation	Class of Recreation Opportunity Spectrum (Cortinovis et al. 2018)	Class combining the highest levels of recreation potential and proximity
Cooling	Cooling capacity class (Zardo et al. 2017)	Cooling capacity class A or B (range: from A to E)

Risk mitigation includes various functions through which ecosystems contribute to buffering hazardous events generated by hydrogeological conditions. Three events were selected for the territory of the municipality: rock falls, landslides, and floods (Table 12.2). For the first two, the most important role in terms of ecosystem services is played by forests. Forests can slow down or prevent rock falls, contributing to the protection of sensitive targets, in particular in the proximity of settlements. The presence of forest also contributes to the prevention and mitigation of landslides through the stabilization of land and the reduction in rainwater runoff. As far as river flooding is concerned, the presence of permeable areas and storage capacity reduces the hazard for downstream areas.

The mitigation of disturbances generated by transport infrastructures includes two regulating ecosystem services: noise mitigation and air purification. With regard to noise mitigation, vegetation can limit the spread of noise generated by vehicular and railway traffic when it is placed between the source and the receptors, so as to constitute a continuous and dense shield. The effectiveness of the shielding depends on many variables, including the thickness of the vegetation layer, the density of vegetation, and the noise frequencies (Van Renterghem et al. 2015). Hence, a detailed evaluation requires consideration of all these aspects. For the purpose of identifying hotspots, wooded areas were considered which, given their location, can act as a noise barrier between transport infrastructures and residential areas.

Regarding air purification, vegetation absorbs gaseous pollutants and intercepts particulate by allowing deposition on its surface (Nowak et al. 2006). Although these functions are also carried out by herbaceous and shrubby vegetation, trees provide the most important contribution. The reduction in air pollutants depends linearly on their concentration in the atmosphere; hence, vegetation located in areas most exposed to pollution provides a larger ecosystem service. Vegetation can also act as a physical barrier, if placed near the source of pollution and to protect sensitive targets. For the purpose of identifying hotspots, wooded areas close to the most heavily trafficked roads and close to the residential areas were considered (Table 12.2).

Forest landscape value refers to one of the many intangible benefits produced by ecosystems, that is, the aesthetic and identity value for local communities. In the case of Trento, this role of the forest areas is particularly important. Understanding which features contribute to defining landscape value is a complex issue. However, the topic has already been the subject of a specific study carried out in the municipal area, which considered the type of forest cover and its environmental value and rarity, also including considerations on the aesthetic value and perception by people.

Food production was selected for its relevance in the territory of the municipality of Trento, in absolute terms the municipality with the largest cultivated area of the Province of Trento. Given the presence of valuable crops such as apple orchards and vineyards, this ecosystem service represents a fundamental economic contribution for the territory. The maintenance of agricultural production is also guaranteed by protecting the most valuable soils from urban growth. The most suitable soils were identified by considering both existing crop types and their suitability to host agricultural activity.

Nature-based recreation has been one of the key aspects addressed during the last years by the city administration through the creation of new peri-urban parks. However, the city is characterized by the proximity of residential areas to different typologies of green spaces where citizens perform a variety of day-to-day recreational activities, from hiking to mountain biking. This makes it difficult to assess the level of provision of recreation opportunities across the city using common indicators such as per-capita green areas. To assess nature-based recreation opportunities, we applied an adjusted version of the ESTIMAP-recreation model (Paracchini et al. 2014; Zulian et al. 2018) using input obtained from a process of expert consultation. Seventeen experts, including officers from several municipal and provincial departments, researchers from various institutions, and local practitioners, were involved through an online questionnaire, and a follow-up discussion, with the purpose of selecting and scoring the most relevant elements that promote or support recreational activities in the local context. The result is a map of the so-called

"Recreation Opportunity Spectrum", which combines information on the recreation potential of the different areas based on the natural elements that are present, and on the proximity, defined as the availability of infrastructures and facilities to access and to use the areas. More details about the model and its application to Trento can be found in Cortinovis et al. (2018) and Cortinovis and Geneletti (2018). Areas classified in the best class of both recreation potential and proximity were included as hotspot for recreation.

Microclimate regulation (cooling) refers to the ability of vegetation to reduce temperatures, thanks to the combined effect of several functions: evapotranspiration, shading, and influence on wind direction and intensity (possible effects of barrier or creation of corridors). The theme is particularly relevant for the valley-floor sector of the city of Trento, where summer temperatures are high due to the combined effect of increasingly intense and frequent heat waves and of the urban heat island. The analysis therefore focused on this area of the city, where the urban and peri-urban green areas contribute to reduce temperature, acting as "cool shelters" and producing a cooling effect also on the surroundings areas (Geneletti et al. 2016; Cortinovis and Geneletti 2018). In this study, we adopted a method to assess the cooling effect specifically designed to support planning decisions at the urban scale (Zardo et al. 2017). The method accounts for the two main ecosystem functions involved in microclimate regulation, that is, shading and evapotranspiration, and assesses them based on three properties of green areas: soil cover, canopy, and size. Once a green area component is classified according to these properties, the model provides the corresponding cooling capacity score, depending on the climatic region of the study area (see Zardo et al. (2017) and Geneletti et al. (2020) for further details).

Two illustrative examples of the ecosystem services maps are presented.

Figure 12.2 shows the Recreation Opportunity Spectrum (ROS) in the city of Trento. The indicator is obtained from a cross-tabulation between the recreation potential, which depends on the intrinsic natural features of the areas, and the availability of infrastructures and facilities that make it possible to use green areas for recreational purposes. The valley floor is mostly characterized by low potential, but high availability of infrastructures and facilities, which makes the existing green areas (urban parks and riverbanks) highly accessible and usable. Outside the urban area, many forests fall in the highest class of the opportunity spectrum, due to the high density of hiking trails and facilities dedicated to sport activities.

Figure 12.3 shows the map of the cooling capacity produced by green infrastructure in the most urbanized area of the city of Trento, that is, the valley floor. The highest classes of cooling effect prevail, due to the presence of close-by forests and of the Adige River and its tributaries that contribute to mitigate the negative effects of the urban heat island and to lower the temperature during summer heat waves. The most disadvantaged areas are in the dense neighborhoods close to the city center and in the northern suburbs, where scarcity of green infrastructure and high rate of soil sealing are observed.

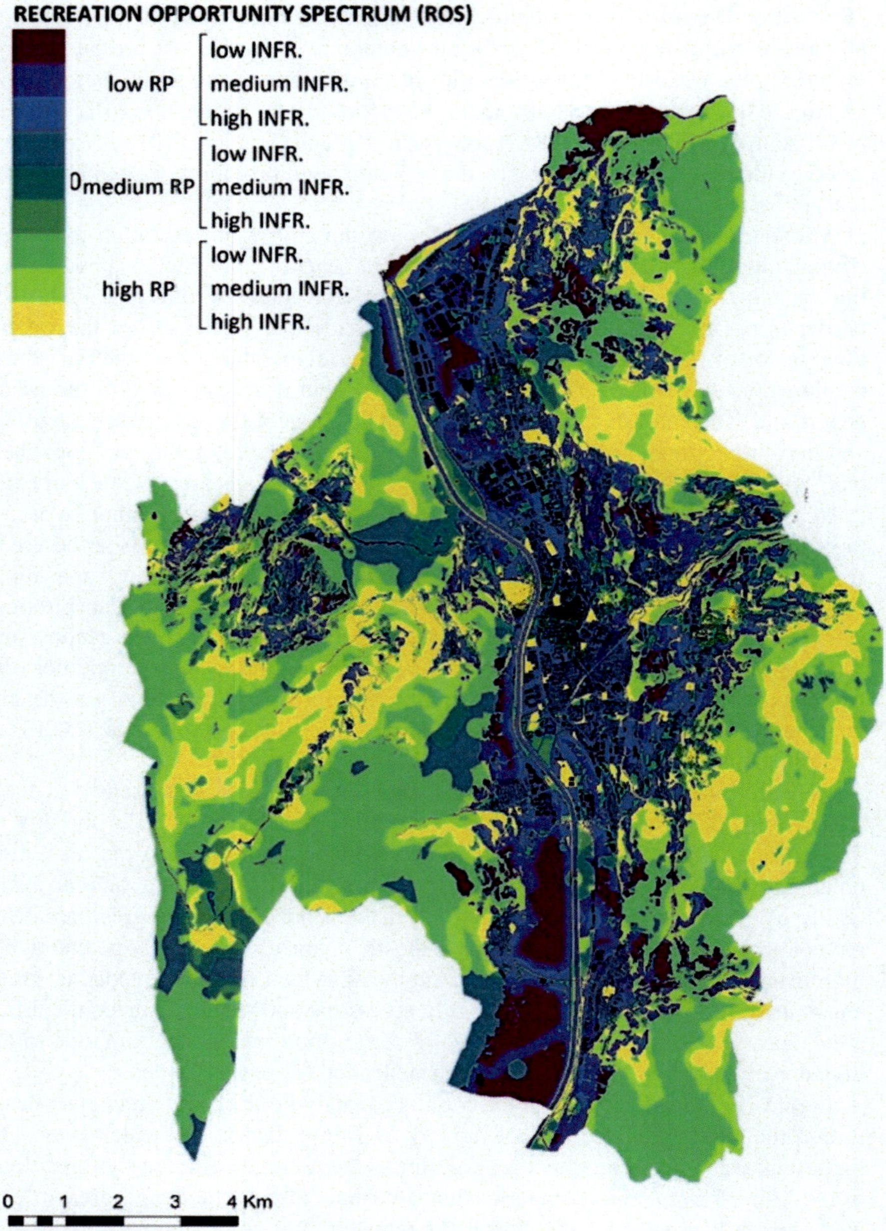

Fig. 12.2 Map of the recreation opportunity spectrum (modified after Cortinovis et al. 2018)

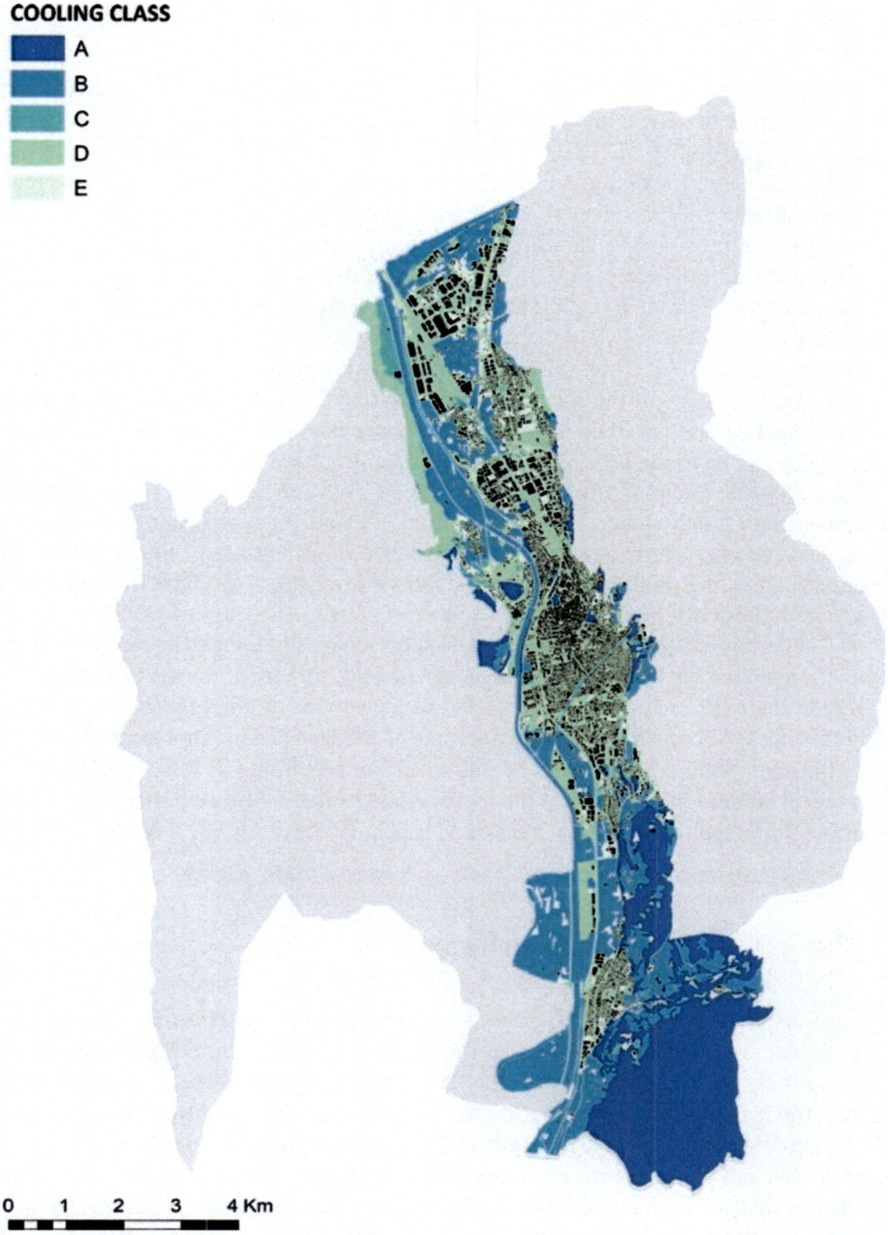

Fig. 12.3 Map of the cooling capacity (modified after Geneletti et al. 2016)

12.3 Mapping Hotspots of Multiple Ecosystem Services

Approaches to identify hotspots of ecosystem services include the selection of a fixed number of cells with the highest levels of ecosystem service supply (e.g., top 20%) (Eigenbrod et al. 2010; Bai et al. 2011), the identification of biophysical thresholds (e.g., for biomass or temperature reduction), and clustering (e.g., Jenks natural breaks) (O'Farrell et al. 2010). Previous studies have identified areas important for the provision of multiple ecosystem services by overlapping individual hotspots (García-Nieto et al. 2013; Peña et al. 2018) or by applying more complex measures of occurrence (e.g., intensity, richness, etc.) (Plieninger et al. 2013).

In this study, we overlaid the seven maps of the hotspots for the individual ecosystem services (see Table 12.2, third column) to obtain a multiple-hotspot map (Fig. 12.4). Values in the map range between zero and five because there are no locations that are classified as hotspots for more than five ecosystem services. The hotspots can be found not only among forests and natural areas surrounding the city, but also in the intensely urbanized valley floor. They are represented by agricultural patches and green wedges, which are often relevant for more than one ecosystem service and characterized by a high biodiversity value. Currently, unlike forests and natural areas that benefit from some levels of protection, the importance of these hotspots is not acknowledged.

The multiple-hotspot map (Fig. 12.4) shows that only approximately 22% of the municipal area has not been classified as a hotspot for any of the seven themes analyzed. These are the urbanized areas and some wooded or not-vegetated portions on both slopes. About 31% of the territory has been identified as a hotspot for only one of the ecosystem services considered, while the remaining 47% has emerged as a relevant area for at least two of the topics considered. On the other hand, only 1.5% of the municipal area has been classified as a hotspot for four or five services.

12.4 Discussion and Conclusions

In terms of policy relevance, the ecosystem service hotspots (considered as both areas that provide multiple ecosystem services, and the most important provisioning area for each ecosystem service) are becoming part of the "structural elements" of the Urban Plan currently being drafted, along with more traditionally recognized elements, such as protected areas or areas subject to hydrological risk. The inclusion of ecosystem service hotspots among the structural elements of the Urban Plan ensures that urban green and blue infrastructures are considered as a primary component of the urban system during the design of the Plan. The hotspots will be preserved from urbanization and different actions are under consideration to improve the current network of green and blue spaces, thus increasing both connectivity and the provision of ecosystem services.

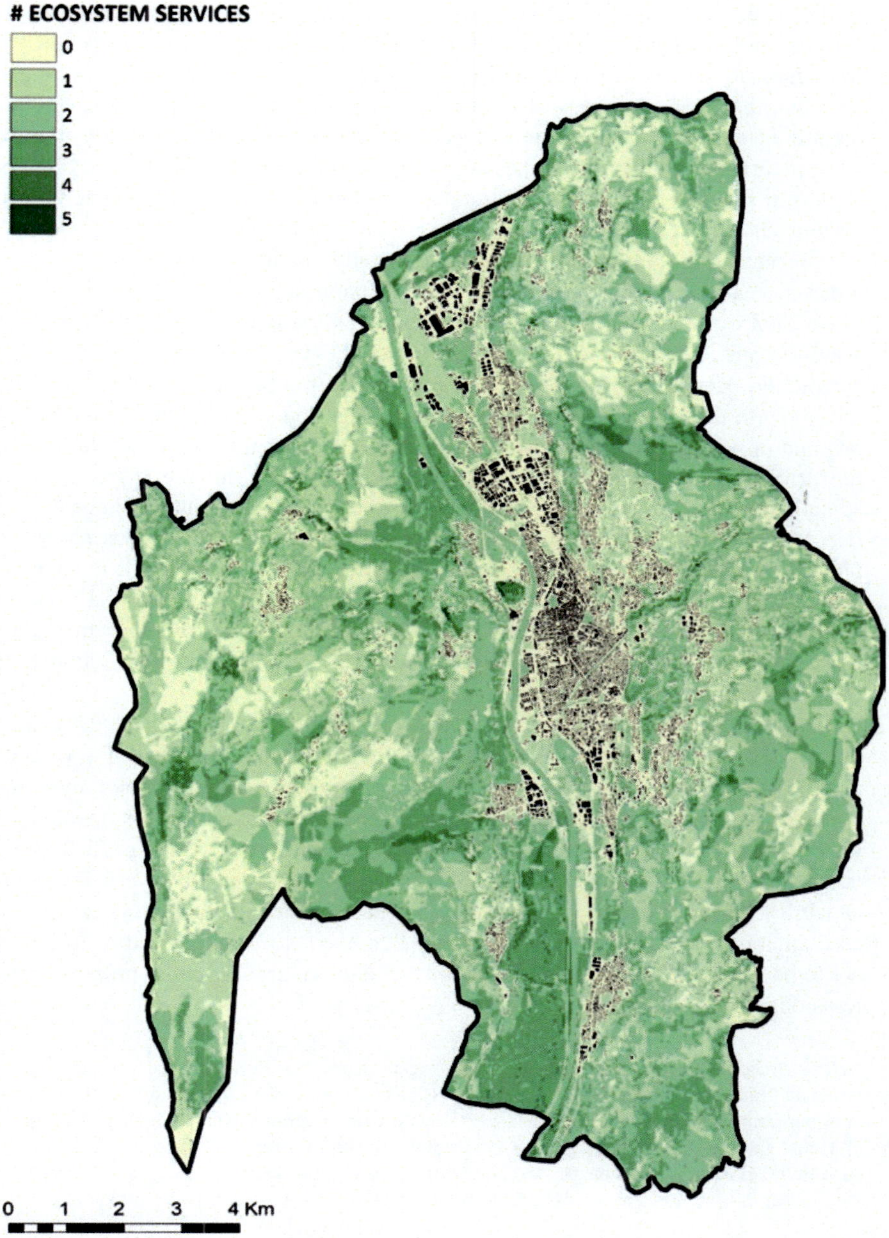

Fig. 12.4 Map of the hotspots for multiple ecosystem services

The results of the hotspot mapping can also be used as a tool for evaluating the actions envisaged by the plan and for monitoring the impacts deriving from its implementation. For example, during the Strategic Environmental Assessment of the plan, the effects in terms of reducing the hotspot areas could be predicted and quantified, allowing to understand which ecosystem services will be mostly affected. This, in turn, would allow to make suggestions to improve planning actions, but also to design suitable mitigation and compensation measures targeted at specific ecosystem services (Tallis et al. 2016).

Concerning the limitations of the present study, the identification of the ecosystem services hotspots has been performed by combining maps with different resolutions. This was due to different indicators and methods used to quantify ecosystem services, but also to the different themes and ecosystem services that have been considered. Hence, the spatial scale of the final output is constrained by the information with the lowest resolution. The resulting outputs are suitable to answer the original policy question concerning the identification of the structural elements of the Urban Plan. However, those outputs are not necessarily suitable, in terms of both spatial and thematic resolutions, to address other policy questions for which ecosystem service knowledge could be equally relevant, for example, the comparison of planning options (see, e.g., Geneletti et al. (2020), p. 47). These might involve elements of the zoning scheme that need to be defined with higher spatial accuracy than that of our ecosystem service maps. Hence, further and more detailed input maps, and modeling approaches, might need to be used. However, this may hamper the future maintenance and use of the results.

In this study, we mainly employed existing maps and data collected for other purposes by the city administration, which could serve as proxies of ecosystem services. While this approach may be less reliable than more complex modeling approaches, a positive aspect is that it uses data already validated and updated. On the contrary, models such as those for the assessment of the cooling effect and of nature-based recreation could not be validated with observed data (yet). The actual capability of the city administration to produce updates of the ecosystem service assessment over the next years is an issue that needs to be addressed if the results are to be used to guide the implementation of the Urban Plan and to propose future revisions.

Acknowledgments The research received support from the European Commission project EnRoute "Enhancing resilience of urban ecosystems through green infrastructure" (https://oppla.eu/groups/enroute), from the "Trento Urban Transformation" project (funded by the Municipality of Trento and the University of Trento) and from the Italian Ministry of Education, University and Research (MIUR) in the frame of the "Departments of Excellence" grant L. 232/2016. We thank the Science Museum of Trento (MUSE) for making available the species distribution data.

References

Bai Y, Zhuang C, Ouyang Z et al (2011) Spatial characteristics between biodiversity and ecosystem services in a human-dominated watershed. Ecol Complex 8:177–183. https://doi.org/10.1016/j.ecocom.2011.01.007

Cortinovis C, Geneletti D (2018) Mapping and assessing ecosystem services to support urban planning: a case study on brownfield regeneration in Trento, Italy. One Ecosyst 3:e25477. https://doi.org/10.3897/oneeco.3.e25477

Cortinovis C, Zulian G, Geneletti D (2018) Assessing nature-based recreation to support urban green infrastructure planning in Trento (Italy). Land 7:112. https://doi.org/10.3390/land7040112

Eigenbrod F, Armsworth PR, Anderson BJ et al (2010) The impact of proxy-based methods on mapping the distribution of ecosystem services. J Appl Ecol 47:377–385. https://doi.org/10.1111/j.1365-2664.2010.01777.x

García-Nieto AP, Garcìa-Llorente M, Iniesta-Arandia I, Martín-López B (2013) Mapping forest ecosystem services: from providing units to beneficiaries. Ecosyst Serv 4:126–138. https://doi.org/10.1016/j.ecoser.2013.03.003

Geneletti D, Zardo L, Cortinovis C (2016) Promoting nature-based solutions for climate adaptation in cities through impact assessment. In: Geneletti D (ed) Handbook on biodiversity and ecosystem services in impact assessment. Edward Elgar Publishing, Cheltenham, pp 428–452

Geneletti D, Scolozzi R, Adem Esmail B (2018) Assessing ecosystem services tradeoffs across agricultural landscapes in a mountain region. Int J Biodivers Sci Ecosyst Serv Manag 14:1–35. https://doi.org/10.1080/21513732.2018.1526214

Geneletti D, Cortinovis C, Zardo L, Esmail BA (2020) Planning for ecosystem services in cities. Springer International Publishing, Cham

Nowak DJ, Crane DE, Stevens JC (2006) Air pollution removal by urban trees and shrubs in the United States. Urban For Urban Green 4:115–123. https://doi.org/10.1016/j.ufug.2006.01.007

O'Farrell PJ, Reyers B, Le Maitre DC et al (2010) Multi-functional landscapes in semi arid environments: implications for biodiversity and ecosystem services. Landsc Ecol 25:1231–1246. https://doi.org/10.1007/s10980-010-9495-9

Paracchini ML, Zulian G, Kopperoinen L et al (2014) Mapping cultural ecosystem services: a framework to assess the potential for outdoor recreation across the EU. Ecol Indic 45:371–385. https://doi.org/10.1016/j.ecolind.2014.04.018

Pedrini P, Tattoni C, Brambilla M (2013) Individuazione della connettivitá e della frammentazione ecologica a livello provinciale e verso i territori limitrofi. Trento

Peña L, Onaindia M, de Manuel BF et al (2018) Analysing the synergies and trade-offs between ecosystem services to reorient land use planning in metropolitan Bilbao (northern Spain). Sustain 10. https://doi.org/10.3390/su10124376

Plieninger T, Dijks S, Oteros-Rozas E, Bieling C (2013) Assessing, mapping, and quantifying cultural ecosystem services at community level. Land Use Policy 33:118–129. https://doi.org/10.1016/j.landusepol.2012.12.013

Tallis H, Kennedy CM, Ruckelshaus M et al (2016) Mitigation for the people: an ecosystem services framework. In: Geneletti D (ed) Handbook on biodiversity and ecosystem services in impact assessment. Edward Elgar Publishing, Cheltenham

Van Renterghem T, Forssén J, Attenborough K et al (2015) Using natural means to reduce surface transport noise during propagation outdoors. Appl Acoust 92:86–101. https://doi.org/10.1016/j.apacoust.2015.01.004

Wolynski A, Masè L, Carriero A, Campana R (2016) La carta dei boschi di protezione da massi della provincia di Trento. Metodologia di elaborazione e modalità di impiego, Trento

Zardo L, Geneletti D, Pérez-Soba M, Van Eupen M (2017) Estimating the cooling capacity of green infrastructures to support urban planning. Ecosyst Serv 26:225–235. https://doi.org/10.1016/j.ecoser.2017.06.016

Zulian G, Stange E, Woods H et al (2018) Practical application of spatial ecosystem service models to aid decision support. Ecosyst Serv 29:465–480. https://doi.org/10.1016/j.ecoser.2017.11.005

Chapter 13
Mapping Ecosystem Services, Disservices, and Ecological Requirements to Enhance Urban Forest Planning and Management in Padova

Chiara Cortinovis, Claudia Alzetta, and Davide Geneletti

Abstract In Padova, the municipal department of green infrastructure monitors the urban forest through a continuously updated database that stores information about locations, species, dimensions, health conditions, and management of more than 47,000 trees. The aim of the research presented in this chapter was to support urban forest planning and management by integrating the existing database with information about ecosystem services, disservices, and the ecological requirements of trees. We combined a quantitative analysis through i-Tree Eco with a qualitative assessment based on literature review and expert knowledge, and filled in 17 new fields for each tree in the database. Then, through a GIS, we investigated the spatial distribution of the analysed features across the city, thus revealing inequalities among different areas and mismatches with citizens' preferences and needs. The enhanced database could potentially become a powerful tool, not only to identify and prioritize management interventions, but also, in a longer term perspective, to identify strategic goals with a view to coordinating local actions, thus ensuring a sustainable development of the urban forest and an equitable provision of ecosystem services to present and future generations.

C. Cortinovis (✉)
Centre for Environmental and Climate Research, Lund University, Lund, Sweden

Department of Civil, Environmental and Mechanical Engineering, University of Trento, Trento, Italy
e-mail: chiara.cortinovis@cec.lu.se

C. Alzetta
Department of Green, Parks and Urban Agriculture, Municipality of Padova, Padova, Italy

D. Geneletti
Department of Civil, Environmental and Mechanical Engineering, University of Trento, Trento, Italy
e-mail: davide.geneletti@unitn.it

A. Arcidiacono, S. Ronchi (eds.), *Ecosystem Services and Green Infrastructure*, Cities and Nature, https://doi.org/10.1007/978-3-030-54345-7_13

Keywords Urban trees · Municipal tree database · Tree management plan · i-Tree Eco · Ecosystem disservices · Tree ecological requirements · Spatial analysis · Urban planning

13.1 Introduction and Objectives

Urban trees are fundamental components of urban green infrastructure and supply cities with key ecosystem services. Trees are especially relevant for the provision of many regulating services, through which green infrastructure affects environmental conditions in cities (Cortinovis and Geneletti 2019). Inter alia, urban trees cool down the temperature during summer through shading and evapotranspiration (Zardo et al. 2017), they purify the air from gaseous pollutants and particulate matter (Morani et al. 2011), they contribute to buffering anthropogenic CO_2 emissions by sequestering and storing atmospheric carbon (Nowak et al. 2013), and they mitigate the propagation of noise (Van Renterghem 2014). In cities, trees also enhance the aesthetic pleasantness of the urban environment, sometimes becoming fundamental presences for the surrounding communities, promoting a sense of place, and strengthening the local identity (Goodness et al. 2016). Furthermore, urban trees support biodiversity and pollination, and can be selected and managed to become a source of food and biomass (Escobedo et al. 2011; Dobbs et al. 2014; Somme et al. 2016).

The term "urban forest" collectively refers to all trees in a certain urban area, including both woodlands or small forest patches and single trees (Nowak et al. 2008; Dobbs et al. 2011; Endreny 2018). However, key differences exist in the capacity of different trees to provide ecosystem services. Depending on species, dimension, age, health conditions, and location, trees substantially differ in the ecological functions they perform, hence in the ecosystem services that they are able to produce (Nowak et al. 2008; Grote et al. 2016). Consequently, the composition of the urban forest determines its capacity to provide ecosystem services.

As living organisms, urban trees also require specific conditions to survive and to grow (Vogt et al. 2017). The harsh urban environment, often characterized by pollution, water scarcity, limited space for roots and crowns, and extreme sun or shade exposure, is not the optimal habitat for many species (Mullaney et al. 2015). Hence, species and varieties have been selected to fit specific urban environmental conditions in different climatic regions across the world (e.g. Sæbø et al. 2003). However, conflicts between trees and humans, sometimes referred to as "ecosystem disservices," are common in cities (von Döhren and Haase 2015). Ecosystem disservices produced by trees include the emission of biogenic volatile compounds (ozone precursors), hardscape damages (i.e. damages related to tree roots causing curbs, sidewalks and other hardscape features to lift), underground infrastructure conflicts (utility wires and pipes), safety hazards from tree falls, and the production of allergenic pollen, among others (Escobedo et al. 2011). Most of these conflicts can be avoided, or at least limited, by a careful planning and management of urban forests that accounts for the ecological requirements of trees.

Urban forest managers usually have a good understanding of the specific needs of different species, as well as of the risks and disservices that they can produce (Conway and Vander Vecht 2015; Davies et al. 2017). However, when selecting new trees to be planted, considering all these aspects along with the ecosystem services that trees can provide is much more difficult (Davies et al. 2017; Roy et al. 2017). Crucially, the process of tree selection normally focuses on matching the characteristics of the tree with the specific feature of the location in which it will be planted, with little attention paid to the aggregate results of single decisions (Conway and Vander Vecht 2015). On the contrary, understanding how ecosystem services and disservices of trees are distributed across the city would reveal to what extent they match with the demand and needs of the urban population (Baró et al. 2015; Ortiz and Geneletti 2018) – a fundamental piece of information to guide the future management decisions from a planning-oriented perspective (Geneletti et al. 2020).

With the aim of supporting a more effective planning and management of urban trees and urban forests, our applied research focused on two specific objectives:

1. Integrating knowledge on ecosystem services, disservices, and ecological requirements to support a more complete assessment of urban trees
2. Investigating how the analysed features are spatially distributed across the city, so as to reveal inequalities and mismatches with demand

13.2 Case Study

The case study is Padova, a city of around 210,000 inhabitants in north-eastern Italy, 20 km west of the Venetian lagoon. Padova is the third largest provincial capital of the Veneto region and the most densely populated (2300 inhabitants/km^2). Soil sealing is one of the main issues in Padova, where 49.3% of the municipal territory is urbanized: the highest share in the region and among the highest in Italy (ISPRA 2018). The high rate of impermeable soil worsens the environmental conditions of the city, which is prone to air pollution, heat waves, and urban flooding.

According to municipal data, green areas total around 550 ha, 5.9% of the territory, equivalent to 26.5 m^2/inhabitant. The availability of green spaces is therefore quite high compared to other cities and urban areas, although only 28% of these are parks and public gardens. Among the most important green infrastructure in Padova are the semi-natural areas along the riverbanks that surround the historical centre, which form popular recreational places for the locals. Riverbanks provide an important opportunity for nature-based activities in the inner districts, characterized by a lower availability of green areas compared to peripheral neighbourhoods.

In this context, urban trees contribute with fundamental ecosystem services to citizens' well-being, especially in the most densely built-up areas. The municipality of Padova manages around 47,000 trees and shrubs of more than 250 species, varieties, and cultivars (data in July 2017). For each tree, continuously updated information about location, structural features, and health conditions are stored in a GIS database, together with a record of management interventions.

In 2017–18, the city was involved in EnRoute "Enhancing resilience of urban ecosystems through green infrastructure", a project of the European Commission in the framework of the EU Biodiversity Strategy and the Green Infrastructure Strategy (Maes et al. 2019). The aim of EnRoute was to provide evidence on how scientific knowledge of urban ecosystems can support urban planning at different stages of policy and for various spatial scales. A city-lab was set up involving key staff from the municipal department of green infrastructure, responsible for managing urban trees and the tree database, and research partners from the University of Trento. Students from the University of Padova working as trainees in the department were also involved in the activities and contributed to data collection and analysis.

The city-lab focused on enhancing the urban tree database created by the municipal department, to boost and exploit its full potential. The overall purpose was to broaden its use from a tool for informing and directing management interventions to a planning support tool that can help to prioritize new planting sites and to select the most suitable species to be used in different areas of the city.

13.3 Enhancing the Urban Tree Database

The methods adopted to enhance the existing database consisted of two main steps: (i) assessing a set of relevant ecosystem services, disservices, and ecological requirements for each tree included in the database; and (ii) conducting a spatial analysis of the data to understand distribution across the city (Fig. 13.1).

13.3.1 Assessing Ecosystem Services, Disservices, and Ecological Requirements of Urban Trees

The assessment of ecosystem services, disservices, and ecological requirements of urban trees was conducted by combining quantitative and qualitative indicators, based on available data recorded in the existing database.

Quantitative assessments were carried out for the provision of relevant ecosystem services and disservices using some of the models available in i-Tree Eco (https://www.itreetools.org/eco/index.php). The analysis was conducted with meteorological and air pollution data for the year 2013. Input data retrieved from the database included species (a total of 255, including varieties and cultivars), dimensions (three categories of diameter at breast height – DBH – and four categories of tree height), and location (street tree vs tree in urban green areas).

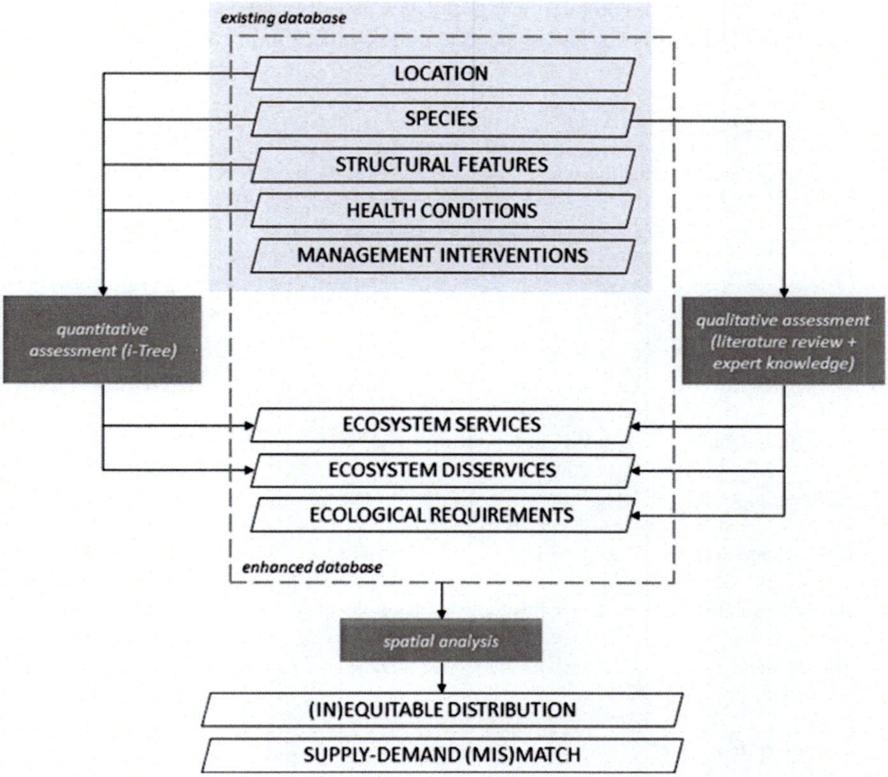

Fig. 13.1 Workflow of the analyses

The results of the quantitative assessment were complemented by qualitative analyses. Based on previous studies, among the information available in the database, tree species were identified as the main characteristic that could be linked to the provision of ecosystem services and disservices, as well as to some ecological and maintenance requirements (Smith et al. 2017; Vogt et al. 2017).

A qualitative scale was defined for each aspect to be assessed, then species were classified according to this scale through a review of existing literature, including databases such as Citree (Vogt et al. 2017), scientific studies on trees in urban areas (e.g. Roloff et al. 2009), and published lists of species associated with specific issues (e.g. toxicity and allergenic potential).[1] In this phase, the knowledge and experience of the municipal staff proved essential to assess the transferability of data from existing studies to the context of Padova, and to complete the assessment for missing species.

[1] See, for example, the list of toxic plants produced by the University of California (https://ucanr.edu/sites/poisonous_safe_plants/Toxic_Plants_by_Scientific_Name_685/) and the reference values for pollen allergenicity determined in Italy (http://www.pollnet.it/valori_di_riferimento_it.asp).

Table 13.1 provides an overview of the new aspects added in the database, including a list of indicators and the respective measurement units used in the assessment.

Table 13.1 Overview of the ecosystem services, disservices, and ecological requirement of urban trees considered in the analysis, the respective methods, indicators, and measurement units. The use of italics indicates that the assessment for all species in Padova is still ongoing

Feature analysed	Indicator	Measurement unit/ scale	Method
Ecosystem services			
Air purification	Pollution removal (NO_2, SO_2, O_3, CO, $PM_{2.5}$)	Tonnes/year	i-Tree Eco
Climate regulation	Carbon storage	Tonnes	i-Tree Eco
	Gross carbon sequestration	Tonnes/year	i-Tree Eco
Run-off mitigation	Avoided run-off due to interception	m^3/year	i-Tree Eco
Pollination (support to)	Honey plant	Yes/no	Literature review + expert opinion
Habitat support	Bird feeding	Yes/no	Literature review + expert opinion
Aesthetic quality	Ornamental value	Low/medium/high	Literature review + expert opinion
Microclimate regulation	*Cooling potential*	*Low/medium/high*	*Literature review + expert opinion*
Ecosystem disservices			
Air quality	VOC emissions	kg/year	i-Tree Eco
Odour nuisance	Bad smell	Yes/no	Literature review + expert opinion
Economic loss	Potential damage by root	Yes/no	Literature review + expert opinion
Safety risk	Stability risk	Low/medium/high/ very high	Literature review + expert opinion
Health risk	*Allergenic potential*	*Low/medium/high*	*Literature review + expert opinion*
	Toxicity	*Low/medium/high*	*Literature review + expert opinion*
Ecological and maintenance requirements			
Overall maintenance requirements	Total costs	Low/medium/high	Literature review + expert opinion
Tolerance to urban environment	*Tolerance to drought*	*Low/medium/high*	*Literature review + expert opinion*
	Tolerance to cold and ice	*Low/medium/high*	*Literature review + expert opinion*

13.3.2 Analysing Spatial Distribution Across the City

Since the database includes information about the geographical location of each tree, it was possible to link the data about ecosystem services, disservices, and ecological requirements generated in the first stage of the analysis to the specific place in the city where the respective tree is located.

Then, a GIS software was used to investigate the spatial distribution of the data. Beyond aggregate results at the city scale, we considered three different subdivisions of the sample relevant for planning and management purposes, based on three different classifications of tree location (Fig. 13.2):

1. Street trees vs trees located in public green areas
2. District: comparing the six main districts of the city, not only in terms of absolute number of trees and tree density, but also in terms of relative performance related to the selected ecosystem services and disservices, and maintenance requirements
3. Distance from the city centre: dividing the map into concentric buffers at a distance of 500 m to capture variations in the presence of the analysed features along a gradient from the centre to peripheral neighbourhoods (the gradient analysis was considered appropriate given the clear monocentric urban structure that characterizes the city)

Fig. 13.2 Map of the city of Padova showing the six main districts and the buffers used for the gradient analysis. On the right: a zoom showing a spatial representation of the tree database

13.4 Ecosystem Services, Disservices, and Ecological Requirements of Urban Trees in Padova

Street trees in Padova represent 24% of the total and are mostly located along the main roads that depart from the city centre and in those neighbourhoods where the urban morphology is traditionally characterized by the presence of tree-lined streets. As a consequence, the share of street trees is higher close to the centre and lower in the peripheral districts, where trees located in public green areas prevail.

The analysis conducted through i-Tree Eco produced some key figures about the ecosystem services provided by urban trees in Padova. Overall, they store 6554 tonnes of carbon, with a sequestration rate of 307.1 tonnes per year. During the reference year, public trees removed around 10 metric tonnes of air pollution, including 6.4 tonnes of ozone (O_3), 2.5 tonnes of nitrogen dioxide (NO_2), 0.7 tonnes of particulate matter less than 2.5 microns ($PM_{2.5}$), and 0.3 tonnes of sulphur dioxide (SO_2), while emissions of volatile organic compounds (precursors to ozone formation) were estimated to be around 3.9 tonnes. Trees also contribute to avoiding surface run-off by intercepting storm water and increasing infiltration and storage in the soil. In 2013, in Padova, the total run-off avoided through the presence of public trees was 25,700 cubic metres.

Figure 13.3 shows two illustrative examples of the results that can be extracted from the spatial analysis of the enhanced database.

The graph in Fig. 13.3a compares the performance of the six main districts of Padova in terms of carbon sequestration, one of the ecosystem services quantified for each tree through i-Tree Eco. The absolute number of trees located in the districts varies from less than 4000 trees in the city centre to more than 12,000 trees in the eastern district. However, the capacity of urban trees to provide ecosystem services depends on a number of features that include their species, age, dimension, and health conditions. These affect the performance of the urban forest in different districts. For example, a comparison between the city centre and the northern district reveals that, despite a lower number of trees, the urban forest in the city centre overall sequesters almost 5 tonnes/year more carbon than that in the northern district.

Figure 13.3b presents the results of the analysis of another ecosystem service included in the enhanced version of the database: the support to pollinators provided by the presence of honey plants. In this case, a gradient analysis was conducted to investigate the performance of different areas of the city, from the centre to peripheral neighbourhoods. The graph shows the number of trees in each buffer and the density of honey plants (which accounts for the progressive increase of the buffer area while moving away from the city centre). Due to a combined effect of tree density and species selection, the density of honey plants shows a peak between 1500 and 2000 m from the city centre, then it drops when it reaches 4500 m. The result points to a scarce provision in peripheral districts, where the presence of honey plants could be useful to support honeybees and pollinators in the proximity of agricultural areas.

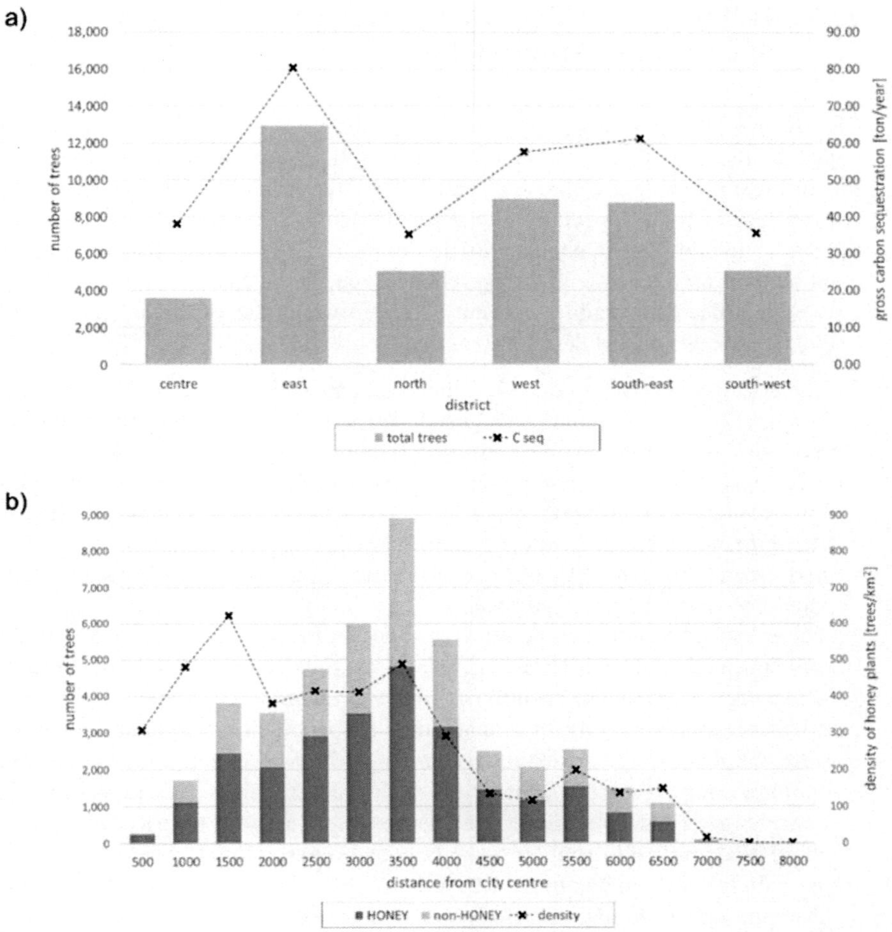

Fig. 13.3 Two illustrative results from the spatial analysis of the enhanced tree database: (**a**) number of trees in each district (left axis) compared with gross carbon sequestration (right axis); (**b**) presence (left axis) and density (right axis) of honey plants along a gradient from the city centre to peripheral neighbourhoods

Other aggregate results reveal interesting and sometimes unexpected consequences of how trees have been selected and managed in Padova during the last decades. For example, trees in urban green areas have on average a higher ornamental value compared to those along the streets, but the percentage of trees providing food for birds is significantly lower (42% vs 61%). Also, contrary to expectations, street trees are characterized by a higher stability risk and a higher percentage of species that are known to produce damages from roots.

13.5 Padova Urban Tree Database: From a Management to a Planning Support Tool

The overall objective of this study was to enhance the database of public trees owned by the city of Padova and to test its potential for supporting decisions about urban forest planning and management. Indeed, the results of the analysis provide relevant information, especially to compare different areas and districts and, more in general, to highlight an unequal distribution of benefits, disservices, and management burdens associated with the presence of trees in the city.

We identified four main opportunity areas in which use of the enhanced urban tree database can support decision-making:

1. The aggregate results produced by the spatial analysis of the enhanced tree database made it possible to identify disadvantaged areas characterized by the scarcity of trees in general, by a comparatively low provision of key ecosystem services, or by a comparatively high intensity of disservices and risks produced by urban trees. Such areas could be targeted by future interventions aimed at strengthening the supply of ecosystem services (e.g. achieving a more balanced performance in terms of carbon sequestration) and reducing ecosystem disservices and management requirements.
2. The enhanced database could serve as a tool for the site-specific prioritization of tree species based on a large set of features, thus supporting tree managers in achieving a more balanced provision of ecosystem services and disservices in different parts of the city, or in matching the specific needs and requirements of different areas. In this case, the assessment of the present conditions of the areas could be compared with an analysis of their different levels of ecosystem service demand or sensitivity to disservices. For example, a list of priority honey plants could be compiled to support honeybees and other pollinators in areas close to agricultural fields and community gardens.
3. The qualitative estimation of the ecosystem services provided by existing trees can serve as baseline information to assess the impact of land-use changes and other urban transformations involving modifications of the urban forest.
4. At a more advanced level, the information stored in the enhanced database can support the design of appropriate and site-specific compensation schemes that account for the impacts of the proposed transformations as well as for the actual conditions and needs of the surroundings. For example, in the case of new urban developments, private developers could be asked to choose among a list of species selected by the municipality for each district or zone based on specific priority goals. Compensation schemes could be used both to promote the provision of relevant ecosystem services and to reduce disservices, for example by limiting the concentration of trees with allergenic potential in certain areas of the city.

Considering these potential uses of the results, a key limitation of the assessment is the exclusion of private areas (Daniel et al. 2016). In the case of Padova, it is estimated that the major part of urban green infrastructure is on private land. Some experiments are ongoing to obtain information about such areas, either by directly involving the citizens in providing data or by making use of high-resolution images and other remotely sensed data (e.g. lidar data). This would allow for a more complete picture of the current provision of ecosystem services and disservices across the city, and hence for a better knowledge base to support planning and management interventions.

Regarding the transferability of the proposed approach to other cities, some important factors should be taken into account. Firstly, both i-Tree Eco and the qualitative approach adopted in the analysis require high-quality and updated data that may not be available in all cities (McPherson et al. 2017). While the complementary weather and pollution data used to run i-Tree is now monitored almost everywhere in Europe, complete databases of public trees such as the one available in Padova are rare. To compile and update such databases, protocols for data collection and management must be included in the routine work of the departments responsible for green infrastructure.

Secondly, even in cases where a complete tree database is available, its enrichment with data about ecosystem service, disservices, and ecological requirements should be a careful process. In the case of Padova, the assessment, revision, and adjustment of classifications found in scientific literature accounting for the specific conditions of the city was a key step in the analysis, made possible only thanks to the knowledge provided by local experts. The resulting classification of species is only partially transferable to other contexts. The same limitation also applies to the use of i-Tree Eco, where many variables must be carefully tuned to the specific context of application (Nowak et al. 2013).

13.6 Conclusions

The experience of the city of Padova presented in this chapter demonstrated how ecosystem service assessments can become a valuable support tool for better management of urban forests, contributing to widening the perspective of tree selection procedures from an exclusive focus on the tree–site relation to a more goal-oriented approach. The enhanced version of the tree database that was compiled by adding to existing monitoring data further information about ecosystem services, disservices, and ecological requirements of all public trees made it possible to analyse how those features are distributed across the city, revealing inequalities and mismatches with the demand arising from citizens and human activities.

The city of Padova is now drafting a Public Tree Management Plan, which will constitute the reference document for planning and managing the city's urban forest in the coming years. There are a number of drivers that will have an impact on Padova's green infrastructure, including climate change, urban growth pressures, and shifts in the needs and preferences of citizens, and all of these deserve attention in the new plan. The municipal department of green infrastructure has been monitoring Padova's urban forest for several years to ensure its health and sustainable management. The enhanced version of the database produced during EnRoute has integrated this knowledge, allowing for careful tree selection that considers the specific purpose of tree planting along with a large set of ecosystem services, disservices, and ecological requirements that should match the needs and characteristics of the planting site and its surroundings. In the context of the new Public Tree Management Plan, the database could be used to create scenarios that account for future changes in both the urban forest and its context, thus identifying long-term policies aimed at a sustainable development of the urban forest and an equitable provision of ecosystem services to present and future generations.

Acknowledgements We are grateful for the assistance of Federico Agostini, who contributed to the analyses as part of his traineeship at the Department of Green, Parks and Urban Agriculture of the municipality of Padova. The research received support from the European Commission project EnRoute "Enhancing resilience of urban ecosystems through green infrastructure" (https://oppla.eu/groups/enroute) and from the Italian Ministry of Education, University and Research (MIUR) in the framework of the "Departments of Excellence" grant L. 232/2016.

References

Baró F, Haase D, Gómez-Baggethun E, Frantzeskaki N (2015) Mismatches between ecosystem services supply and demand in urban areas: a quantitative assessment in five European cities. Ecol Indic 55:146–158. https://doi.org/10.1016/j.ecolind.2015.03.013

Conway TM, Vander Vecht J (2015) Growing a diverse urban forest: species selection decisions by practitioners planting and supplying trees. Landsc Urban Plan 138:1–10. https://doi.org/10.1016/j.landurbplan.2015.01.007

Cortinovis C, Geneletti D (2019) A framework to explore the effects of urban planning decisions on regulating ecosystem services in cities. Ecosyst Serv 38:100946. https://doi.org/10.1016/j.ecoser.2019.100946

Daniel C, Morrison TH, Phinn S (2016) The governance of private residential land in cities and spatial effects on tree cover. Environ Sci Pol 62:79–89. https://doi.org/10.1016/j.envsci.2016.01.015

Davies HJ, Doick KJ, Hudson MD, Schreckenberg K (2017) Challenges for tree officers to enhance the provision of regulating ecosystem services from urban forests. Environ Res 156:97–107. https://doi.org/10.1016/j.envres.2017.03.020

Dobbs C, Escobedo FJ, Zipperer WC (2011) A framework for developing urban forest ecosystem services and goods indicators. Landsc Urban Plan 99:196–206. https://doi.org/10.1016/j.landurbplan.2010.11.004

Dobbs C, Kendal D, Nitschke CR (2014) Multiple ecosystem services and disservices of the urban forest establishing their connections with landscape structure and sociodemographics. Ecol Indic 43:44–55. https://doi.org/10.1016/j.ecolind.2014.02.007

Endreny TA (2018) Strategically growing the urban forest will improve our world. Nat Commun 9:10–12. https://doi.org/10.1038/s41467-018-03622-0

Escobedo FJ, Kroeger T, Wagner JE (2011) Urban forests and pollution mitigation: analyzing ecosystem services and disservices. Environ Pollut 159:2078–2087. https://doi.org/10.1016/j.envpol.2011.01.010

Geneletti D, Cortinovis C, Zardo L, Esmail BA (2020) Planning for ecosystem services in cities. Springer International Publishing, Cham

Goodness J, Andersson E, Anderson PML, Elmqvist T (2016) Exploring the links between functional traits and cultural ecosystem services to enhance urban ecosystem management. Ecol Indic 70:597–605. https://doi.org/10.1016/j.ecolind.2016.02.031

Grote R, Samson R, Alonso R et al (2016) Functional traits of urban trees: air pollution mitigation potential. Front Ecol Environ 14:543–550. https://doi.org/10.1002/fee.1426

ISPRA (2018) Consumo di suolo, dinamiche territoriali e servizi ecosistemici. ISPRA, Roma

Maes J, Zulian G, Günther S et al (2019) Enhancing resilience of urban ecosystems through green infrastructure (EnRoute). Final report. Publications Office of the European Union, Luxembourg

McPherson EG, Xiao Q, van Doorn NS et al (2017) The structure, function and value of urban forests in California communities. Urban For Urban Green 28:43–53. https://doi.org/10.1016/j.ufug.2017.09.013

Morani A, Nowak DJ, Hirabayashi S, Calfapietra C (2011) How to select the best tree planting locations to enhance air pollution removal in the MillionTreesNYC initiative. Environ Pollut 159:1040–1047. https://doi.org/10.1016/j.envpol.2010.11.022

Mullaney J, Lucke T, Trueman SJ (2015) A review of benefits and challenges in growing street trees in paved urban environments. Landsc Urban Plan 134:157–166. https://doi.org/10.1016/j.landurbplan.2014.10.013

Nowak DJ, Crane DE, Stevens JC et al (2008) A ground-based method of assessing urban Forest structure and ecosystem services. Arboric Urban For 34:347–358. https://doi.org/10.1039/b712015j

Nowak DJ, Greenfield EJ, Hoehn RE, Lapoint E (2013) Carbon storage and sequestration by trees in urban and community areas of the United States. Environ Pollut 178:229–236. https://doi.org/10.1016/j.envpol.2013.03.019

Ortiz MSO, Geneletti D (2018) Assessing mismatches in the provision of urban ecosystem services to support spatial planning: a case study on recreation and food supply in Havana, Cuba. Sustainability 10:2165. https://doi.org/10.3390/su10072165

Roloff A, Korn S, Gillner S (2009) The climate-species-matrix to select tree species for urban habitats considering climate change. Urban For Urban Green 8:295–308. https://doi.org/10.1016/j.ufug.2009.08.002

Roy S, Davison A, Östberg J (2017) Pragmatic factors outweigh ecosystem service goals in street tree selection and planting in South-East Queensland cities. Urban For Urban Green 21:166–174. https://doi.org/10.1016/j.ufug.2016.12.003

Sæbø A, Benedikz T, Randrup TB (2003) Selection of trees for urban forestry in the Nordic countries. Urban For Urban Green 2:101–114. https://doi.org/10.1078/1618-8667-00027

Smith AC, Harrison PA, Pérez Soba M et al (2017) How natural capital delivers ecosystem services: a typology derived from a systematic review. Ecosyst Serv 26:111–126. https://doi.org/10.1016/j.ecoser.2017.06.006

Somme L, Moquet L, Quinet M et al (2016) Food in a row: urban trees offer valuable floral resources to pollinating insects. Urban Ecosyst 19:1149–1161. https://doi.org/10.1007/s11252-016-0555-z

Van Renterghem T (2014) Guidelines for optimizing road traffic noise shielding by non-deep tree belts. Ecol Eng 69:276–286. https://doi.org/10.1016/j.ecoleng.2014.04.029

Vogt J, Gillner S, Hofmann M et al (2017) Citree: a database supporting tree selection for urban areas in temperate climate. Landsc Urban Plan 157:14–25. https://doi.org/10.1016/j.landurbplan.2016.06.005

von Döhren P, Haase D (2015) Ecosystem disservices research: a review of the state of the art with a focus on cities. Ecol Indic 52:490–497. https://doi.org/10.1016/j.ecolind.2014.12.027

Zardo L, Geneletti D, Pérez-Soba M, Van Eupen M (2017) Estimating the cooling capacity of green infrastructures to support urban planning. Ecosyst Serv 26:225–235. https://doi.org/10.1016/j.ecoser.2017.06.016

Chapter 14
Messina. Green and Blue Infrastructures for the Re-urbanisation of the City

Carlo Gasparrini and Anna Terracciano

Abstract The strong criticalities expressed by its environmental condition make Messina a *paradigm city* to face the multiplicity of factors that stress cities in the world today, dangerously intercepting the risks deriving from climate change, to which are added the effects of the economic and social crises. Thus, the new Urban Plan takes on the integrated interpretation of risks as an opportunity in the defining of a way of shaping the resilient metamorphosis of the city, based on adaptive and proactive tactics and design actions that entrust a central role to green and blue infrastructures (GBI). This dense network of GBI, starting from the existing ones, moves within a territorial and local dimension to contrast the conditions of fragility, reduce exposure and vulnerability, and maximise biodiversity and the production of ecosystem services also in urban areas, contributing to greater safeguarding of the areas at risk and the regeneration of the territory. Indeed, although they have a systemic approach, they take shape through places, resources and practices, representing an open network of multi-functional and multi-scale relations for the testing of places with landscape and ecological quality, with inclusive social practices, innovative economies and public–private collaborative processes.

Keywords Climate change · Integrated risks · Resilient metamorphosis · Adaptive and proactive tactics · Green and blue infrastructures · Ecosystem services · Landscape · Inclusive social practices · Circular economy

---

Carlo Gasparrini: Author of paragraphs 1,2 and 3; Anna Terracciano: Author of paragraphs 4 and 5. All the images accompanying this text are elaborations produced during the drafting process of Preliminary outline of the new Messina City Plan and are edited by Anna Terracciano

---

C. Gasparrini (✉) · A. Terracciano
Department of Architecture, University of Naples "Federico II", Naples, Italy
e-mail: carlo.gasparrini@unina.it; anna.terracciano2@unina.it

A. Arcidiacono, S. Ronchi (eds.), *Ecosystem Services and Green Infrastructure*, Cities and Nature, https://doi.org/10.1007/978-3-030-54345-7_14

14.1 Articulation and Interaction of Risks

Messina is characterised by a complex context of risks that presents the profiles of a peculiar exasperation, which can also be found to varying degrees in many other Italian cities:

- It is a medium-sized city (just under 240,000 inhabitants) with a very large municipal area (211 sq km compared to the 180 sq km average for Italian regional capital cities), which includes a large stretch of the Peloritani Mountains with about 70 torrents characterised by a widespread hydrogeological condition of risk. It presents a high level of seismic hazard and was completely razed in 1908 by a catastrophic earthquake and the consequent tsunami.
- It was rebuilt with a plan with a high consumption of soil and low density, which provided for the start of urban growth processes along some torrents, directed towards the hilly slopes, confirmed and amplified by the settlement dynamics of the second half of the twentieth century through a succession of oversized plans, including the existing one, a pervasive consumption of hilly soil and a progressive "cementificazione" (introduction of concrete or building on) of some torrent and river beds.
- These choices have determined a widespread condition of geomorphological, hydrogeological, hydraulic and seismic criticality, triggering a series of landslides, floods and rivers bursting their banks, up to the tragic mudslide in the Giampilieri district in 2009 with 37 deaths. There is therefore an intense cumulation and reciprocal amplification of the size, complexity and extension of the danger, exposure and vulnerability factors with respect to the various risks inherent to the physical safety of the territory and of the human settlements. In addition to this, there is the ineffectiveness, technical backwardness and lack of integration of the risk maps that are made available by the superordinate public bodies starting from the Regional and the River Basin Authorities.
- The increasing risk conditions are amplified by the constructional poverty and seismic vulnerability of new building and intersect with other types of risk closely connected to the characteristics of the process of the building of the city and its metabolism (Wolman 1965) during the twentieth century. Scarcity and vulnerability of water resources, soil consumption and pollution, low level of urban plant resources, environmental pervasiveness of urban traffic and air pollution, microclimatic vulnerability of open spaces, unsustainability of the levels of energy consumption, lack of control of the waste cycle, widespread production of waste and abandoned areas: Messina records and amplifies, in forms that are sometimes extreme, most of these risk conditions that are stressing the cities of the world in this historical phase, which are even more amplified by the climate changes[1] in progress.

[1] See https://unfccc.int/

14.2 Messina: A Resilient and Anti-fragile City

The new urban plan takes on the integrated interpretation of risks as an opportunity to outline a priority strategic objective in defining a resilient metamorphosis of the city, based on adaptive and proactive tactics and design actions that entrust a central role to green and blue infrastructures (GBI) (Figs. 14.1 and 14.2).

The preliminary outline of the Plan (Comune di Messina 2018) has initiated a chain of interpretations of the urban, environmental and social repercussions that these conditions produce, starting from the background noise that is expressed in the daily risks of an urban metabolism that has gone mad, up to the extreme and undisputed peaks of increasingly ordinary calamitous events.

The diversification and integration of strategies and actions related to the primary environmental resources pass primarily through the rethinking and recycling of water and its reach in the city, the proactive contrast to the consumption of soil, the preservation and the increase of the permeability of the soil and new policies for the decontamination of polluted soils, the reduction of sources of pollution and the increase in urban plant resources, the greater efficiency of structural, energy use and facilities of the existing fabric, the recycling of waste and waste areas, the development of soft mobility and the strengthening of public rail transport.

However, this change of priorities, the multi-scale nature of strategies and actions, the ability to integrate them and make them synergic over time and space must serve an idea of the city legitimised by an adequate level of cultural awareness and not entrusted to sectoral urban policies. In this sense, the materials of resilient actions will contribute to providing Messina with a network of environmental infrastructures that are capable of constituting the frame of a city that imagines itself increasingly less pervasive from the construction point of view and increasingly focused on valorising the geography of a territory with exceptional landscape quality.

In this perspective, the construction of a specific "Action Plan" (De Cola and Gasparrini 2017) for Messina, inserted in the first "Report" of the *Casa Italia* Mission Structure of the Presidency of the Council of Ministers, gave a glimpse of a public planning and programmatic perspective within an integrated dimension of risks, both at the local scale of the city of Messina and at a national level, as a paradigmatic example of future policies to reduce Italian cities' exposure to risk. At the same time, the definition of the preliminary outline of the new master plan of the city, starting from this objective of reducing exposure to hydrogeological, hydraulic and seismic risks, intends to prefigure a wider perspective of adaptation within a geostrategic dimension for the whole city.

Messina thus aims to search for possible and compatible forms of adaptation in areas of moderate risk and, in some cases, to withdraw progressively from the torrent and river beds and from areas of high-risk and high environmental sensitivity, favouring densification and compactness and valorising the diversity of the landscape of the different forms of settlement. The construction of an "Integrated charter of risks and susceptibility to urban redevelopment," prefigured as a first

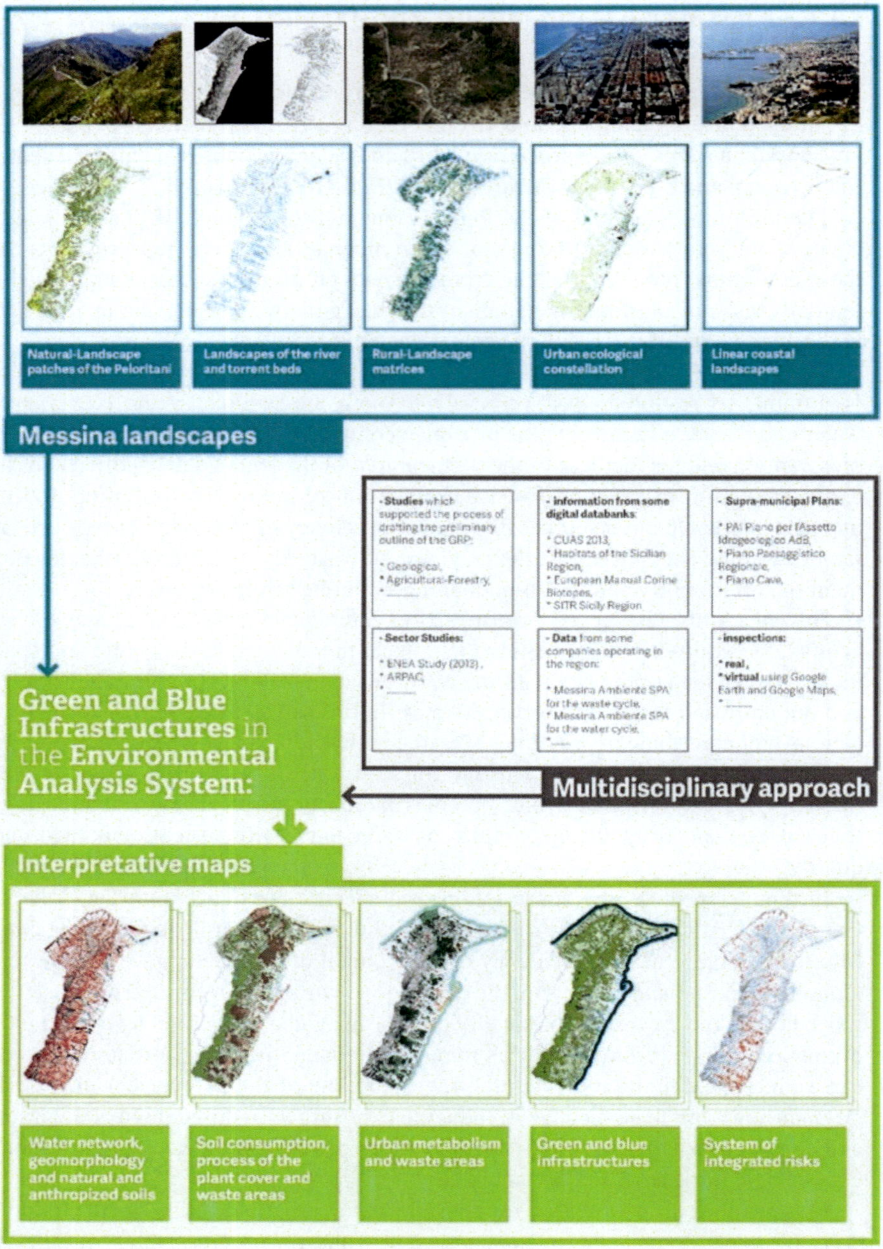

Fig. 14.1 Guidelines for the environmental infrastructures project: general scheme and multi-scalar approach

Fig. 14.2 Green and blue infrastructures: adaptive and proactive tactics and design actions

approximation in the preliminary outline of the plan, may constitute the dynamically updated reference for a strategy of contrast and adaptation to a multiplicity of risks that would be able to accommodate ecological-environmental, social and economic ones. This change describes an idea of a resilient city that valorises the noble parts of urban history (the regular "chessboard" of the Borzì Plan, the network of villages and the historical and architectural emergencies) and finally takes the perverse intersection of existing risks into account – primarily hydrogeological, hydraulic and seismic ones. In this sense, it starts a slow and incremental process of ecologically oriented regeneration to adapt to these critical conditions, raising the performance quality of the existing fabrics and open spaces.

This is a strategy at two scales – regional and local – which mainly relies on the creation of a GBI network, starting from the existing ones, capable of combating fragile conditions and, at the same time, maximising biodiversity and the production of ecosystem services even in urban areas (Sinnett et al. 2015).

14.3 An Incremental Frame for the Resilient Metamorphosis of Messina

The role of the GBIs in this change of direction with respect to the past takes its cue from an evaluation of the most fertile legacy of the debate at the international level in the last 15 years. In the experience of urban, strategic and operational planning of cities and urbanised territories – from large metropolitan areas to intermediate cities up to inland areas – the GBIs tend to take on a central role in the search for possible territorial coalescences and viable prospects for cohesion, which are ecologically oriented, socially inclusive as well as capable of targeting new circular urban economies (Benedict and McMahon 2006). This process is particularly substantial and articulated in Europe where, in the guidelines of urban policies and the most advanced regulatory and managerial European Union guidelines, in the increasingly central role of cities and their networks and alliances on the issue, as well as in the cultural debate and in research, the need for a complex and integrated dimension of the GBIs is being affirmed to respond to the demand for an inescapable resilient re-urbanisation of landscape, environmental and functional quality (Gasparrini 2019).

The virtuous convergence of different questions, to which sectoral and fragmentary answers have always been given, allows us to go beyond the traditional environmental field in which the GBIs have been imagined and intercept more complex urban, social, economic and managerial declinations that are more closely connected to the pervasiveness of the outcomes of the dynamics of contemporary urbanisation and its various ecological, social and economic risks (Beck 2013).

The GBIs are reflected in the extensive production of EU policies and strategies over the last 10 years (Mell 2008, 2015), the 2009 White Paper on adaptation to climate change (European Union 2009) and the European Biodiversity Strategy (European Union 2010) not to mention the Green Infrastructure Strategy of 2013. It is the meaning of these networks that has suffered in the last few years from a substantially environmental perspective, poorly in step with the multidimensional complexity that has been taking shape in European cities.

An explicit and convincing reference for Green Infrastructures can be found in 2017 in the Bologna Charter for the Environment[2] undersigned by the metropolitan cities – Messina included – for sustainable development, following the approval of the 2030 Agenda by the United Nations in 2015 (United Nations 2015): "Recognizing green infrastructures as indispensable elements for climate change adaptation and mitigation, for increasing territorial resilience and for enhancing ecosystem services" (supply or procurement, regulatory, cultural and support services).[3] The earmarking of new European Funds for Regional Development and Cohesion 2021–2027 opens a virtuous perspective in this context to affirm an integrated vision of GBIs to give effective answers to a plurality of questions, integrating economic planning objectives with those of an urban, strategic and operational planning of local administrations. A "Greener Europe" is one of the five strategic objectives underpinning the proposal for the Regulation for the Programming of the new 7-year period, approved by the European Commission.[4] A goal that must be pursued "through the promotion of a transition to clean and fair energy, green and blue investments, the circular economy, adaptation to climate change and risk management and prevention" (European Commission 2018). Therefore, an integrated dimension of the green perspective in our cities and in our territories entrusts the environmental infrastructures with the ability to triangulate different fields of public action, favouring interactions and complementarity with other public and private financial channels within the ideas of a city of which we hope to see an incremental resilient metamorphosis (Gasparrini 2017a, b). In this evolving framework, the ongoing experience at Messina interprets the role of the GBIs at the same time as follows:

- A dynamic and resilient system of adaptation to the multiplicity of environmental risks, amplified by climate changes on planetary and urban scales
- The widespread, growing range of the production of ecosystem services for active and compensating counteractions to the criticalities produced by soil consumption
- The primary reference context for the re-organisation of urban metabolism and the life cycles of resources (primarily water, soil, plant resources, waste, energy)
- The frame of the new public city characterised by high urban and ecological-environmental standards

In this sense, it is possible to include in the GBIs different declinations and fields of public action which, in international experience starting from those in Europe, appear mixed and integrated differently:

---

[2] http://www.comune.bologna.it/sites/default/files/documenti/Carta%20di%20bologna%20per%20l%27ambientepdf

[3] See the definition of the Millennium Ecosystem Assessment https://www.millenniumassessment.org/en/indexhtml

[4] https://ec.europa.eu/regional_policy/it/2021_2027/

- Water networks and technical infrastructures of urban and territorial drainage designed for a dynamic city–water coexistence, mitigation and adaptation to hydrogeological and hydraulic risk, retention and recycling of water resources
- Networks and constellations of vegetative landscapes and permeable soils or soils that need unsealing to guarantee the production of ecosystem services, the replenishment of groundwater, the management of evapotranspiration processes, the reduction of CO2 and climate-altering gases in the atmosphere, the improvement of urban microclimatic conditions (starting from the heat islands), air quality and urban ventilation
- Networks and constellations of drosscapes (polluted soils and bodies of water, abandoned residential or industrial and marginal areas, etc.) to be reclaimed, renaturated and recycled for ecologically oriented collective, social and productive uses
- Networks of roads and underground utilities that are adequate for the current and future demands for space for soft mobility, infrastructures for water retention, disposal and recycling, and energy and digital infrastructures
- Frame of public spaces of landscape quality for the identity, social life and security of the territories and communities
- Places of convergence of actions with a multi-actor approach for social re-appropriation and the creation of collaborative accords, and agreements relating to the management of common goods
- Privileged fields for the development of innovative urban economy production chain linked to recycling and the circular economy, inter-related to the production and management of common goods included in the GBIs (water, soil, greenery, waste, energy, soft mobility, welfare)

This obviously requires a series of choices that put the instrumentation, procedures, rules and organisation forms of the Public Administration under tension. In fact, they call for a convergence and complementarity of public resources at all scales, the activation of urban planning incentives and tax breaks aimed at the objective to match specific "cost centres" at the local level, the push for a new geography of social actors and entrepreneurial and contractual and partnership tools, multi-level governance and an internal re-organisation of public structures according to objective, the re-organisation of the production chains of plans from the vast to the municipal scale to ensure the effectiveness of public action.

14.4 The GBIs to Describe the Landscapes of Messina

The GBIs in Messina are configured like a huge frame that, with different gradients of naturalness, penetrates from the mountains to the sea mainly along the river beds, innervating the anthropic systems in the urbanised areas, also thanks to the strong pressure exerted by the Peloritani and its slopes that extend towards the coast. The GBIs thus pass through all types of landscapes, coming into contact with a territory of disuse and abandonment, empty spaces, environmental and infrastructural criticalities, social conflicts and crises of some production cycles.

The highly critical aspects expressed by its environmental condition constitute one of the distinctive figures of the Messina landscape and make it a *paradigm city* to face the multiplicity of factors[5] that in this historical phase put cities in the world under stress, also dangerously intercepting the risks deriving from the changes in addition to the visible effects of economic and social crises, expressed above all by the lack of equal access to resources and the right to the city (United Nations 2019).

From the reading of the historical territory and from the succession of settlement processes according to some significant temporal analyses, it is possible to deduce the degree of permanence and persistence that the GBIs still preserve as a supporting structure, capable of innervating the entire metropolitan dimension, while profoundly modifying itself through urban, peri-urban and natural contexts. This founding, structuring dimension has enormous potential in terms of regeneration of the GBIs and crosses all the documents in the Environmental System and, more generally, in the new Plan in which the GBIs take on a significant and innovative value because they propose the overcoming of the traditional meaning of an environmental and ecological network, entering into synergy with the landscape in its broadest and most contemporary meaning (LOTUS 1999).

The change[6] in the concept of landscape – understood today as an interpretative, narrative and planning device – towards an integrated and unitary conception, is reaffirmed also in the European Landscape Convention[7] of 2000, extending, in the works of the new City Plan,[8] to the whole of the territory and to the differentiated outcomes of the processes of interaction, stratification and sedimentation of human activities with natural components. These reports allow the identification of a set of territorial identities, also for the purpose of multidisciplinary qualification of landscape-environmental planning, overcoming the idea of landscape as limited to some historical-environmental excellences. The decoding and narration of the features of the Messina landscape was a complex process, characterised by a multiscalar dimension, the result of the interaction and multidisciplinary convergence of different knowledge that has made it possible to build an updated picture of the territorial structure, plant landscapes, evolutionary dynamics of use and consumption of soils, areas of environmental criticalities and, more generally, of the

[5] Of the numerous reports and world events we report: Brundtland Report (1987); Rio de Janeiro Environmental Conference (1992); Agenda 21 (1992); UNEP World Environment Report (GEO 5) (2012); Rio +20 Conference (2012).

[6] Also for the influence of Landscape Planning (McHarg 1997; Mostafavi and Doherty 2010; Waldheim 2006) and Landscape Ecology (Forman and Godron 1986)

[7] The European Landscape Convention defines the landscape as "An area, as perceived by people, whose character is the result of the action and interaction of natural and/or human factors" (Council of Europe 2010).

[8] A selection of the main documents and the report of the Preliminary Urban Design of the City Plan for Messina is available at https://www.comunemessina.gov.it/area-pol-territorio/schema-di-massima-prg-2018/

outcomes of urban metabolism and the more or less virtuous functioning of the city itself.

This process re-establishes a geography of the places where the components of the GBIs that characterise the territory and are easily recognisable on a large scale can mainly be attributed to five categories of landscapes: (1) the "Natural-Landscape patches of the Peloritani," (2) the "Comb-like structure of the river beds," (3) the "Rural-Landscape matrices," (4) the "Urban ecological constellation" and (5) the "Linear coastal landscapes" (Fig. 14.3).

The materials that compose them and the relations they have with each other constitute the structure of the territory itself and are of fundamental importance to understand the relationships that exist between the hydro-geomorphological and vegetation aspects, but also between the consolidated and recently expanding urban settlements with agricultural contexts and coastal areas. This process was based on the interaction of some analytical-specialist readings from different sources such as: (1) Studies (Geological and Agricultural-Forestry) which supported the process of drafting the preliminary outline of the City Plan; (2) information from some digital databanks (SIT Regione Sicilia, CUAS 2013, Charter of the Habitats of the Sicilian Region, Habitat of the European Corine Biotopes Manual, SITR Sicily Region,[9] etc.); the forecasts of the Supra-municipal Plans and those of the Sector; some Sector Studies such as the ENEA Study (2013) for the assessment of landslide hazard in the municipal area; data from some companies operating in the region (such as Messina Ambiente SPA[10] for the waste cycle); real and virtual inspections using Google Earth and Google Maps. The outcomes of this fact-finding process are represented in the interpretative readings contained in five families of papers, amongst which the "Integrated charter of the risks and the susceptibility to urban redevelopment" and the "Urban metabolism and dross spaces"[11] stand out for their newness and innovation.

The "Map of the integrated risks and susceptibility to urban redevelopment" is a multi-risk map and the result of the relationships and overlaps between the map of superordinate constraints, geological hazards and seismic hazards,[12] also considering some active geological conditions that determine changes to the territory (e.g. processes like flooding, landslides, permanent soil deformation due to seismic inputs), and represent a working document that aims to provide a summary picture of the state of vulnerability of the territory[13] to guide the choices of the new Plan towards an integrated, resilient and adaptive mitigation dimension.

The map "Urban metabolism and dross spaces" represents the multiplicity of drosscapes, abandoned buildings, residual spaces, and brownfields intercepted by the continuity of infrastructural networks and by the reticular dimension of

[9] http://www.sitr.regione.sicilia.it/

[10] http://www.messinambiente.it/

[11] See https://www.comunemessina.gov.it/area-pol-territorio/schema-di-massima-prg-2018/

[12] Or Level 1 seismic microzonation (MS1).

[13] See Geological study.

Fig. 14.3 Environmental analysis system: methodological scheme to describe the landscapes of Messina

ecological connections, restoring a critical porous structure that traverses the urban areas at various depths, inevitably overlapping with processes of soil consumption still in progress and the exhaustion of economic, productive and ecosystem life cycles. In Messina, the complex dimension of drosscapes but also and above all the mechanisms and outcomes of an out-of-control urban metabolism can be mainly traced back to the crisis of five life cycles: (1) compromised ecosystems (forestation, coast), (2) production cycles (agriculture, industry), (3) critical fabrics (services and equipment, settlements and residential buildings), (4) compromised infrastructure (mobility, special products) and (5) waste cycle.

14.5 Guidelines for the Environmental Infrastructures Project

From the reading of the territory, the centrality assumed by the environmental components emerges in the direction of a complex multi-system network of GBIs as a new potential resilient re-urbanisation frame for the city because it is not only able to penetrate into urban contexts to deeply regenerate the existing fabrics (Fabos and Ahern 1995), focusing on performance standards of high environmental quality, but also to intercept that constellation of existing micro and macro green pores, not only to reconstruct the ecological continuity from the mountains to the sea through reconnection and reconfiguration operations, but also to redefine a new relationship of coexistence between waters, communities and cities.

The Project Guidelines for Environmental Infrastructures in the Preliminary outline of the new City Plan aim to set up a frame of GBI Networks and Landscapes that plays a role of generator and regenerator for a virtuous urban metabolism in which stability objectives, richness and variety of natural and anthropic ecosystems are guaranteed, avoiding the formation of barriers or solutions of continuity between the habitats concerned. The role of an eco-landscape supporting framework, favouring a resilient, systemic and multi-scalar approach capable of providing adaptive and dynamic solutions, plays an essential role in protecting and strengthening the characteristics of the historicised landscape but also of its vulnerability.

For these reasons, the choices of the new City Plan for the Environmental Infrastructure System were based on the interaction of the following factors: (1) the implementation of guidelines, prescriptions and constraints contained in the superordinate Plans and in national and regional laws in specific sectors and components of the territory; (2) the evaluation of the interpretative outcomes of some specialised analyses prepared in support of the City Plan to represent an updated picture of the territory and of the risk conditions; (3) the evaluation of the interpretative results of some analyses developed to represent the characterising and qualifying environmental components of the Messina landscape; and (4) the results of the first evaluations for the identification of the Urban Regeneration Areas – ARU.[14]

The evaluation of these factors should be considered within a broader framework of priority objectives[15] where the main strategic guidelines identify the safeguarding and valorisation of landscapes, the virtuous adaptation to risk conditions and the awareness of the widespread demand for urban and environmental regeneration as the main fields of action of the choices of the new Plan.

The guidelines for the GBI System Project thus acquire a priority role in support of this strategic framework in which with the prospect of focusing on the regeneration of the existing city and, in particular, on destructured and precarious fabrics, on recycling and the re-use of buildings and of the special abandoned complexes, on the densification of low density and degraded fabrics and on the widespread recovery of unoccupied housing, it perfectly combines with the need not to further increase the consumption of new soil.

In particular, the large strategic fields for multi-scalar and inter-systemic actions are identified by the following landscapes:

- The "Natural-Landscape patches of the Peloritani" for which interventions are envisaged to valorise the woodland landscape as the large core area of a larger metropolitan park that goes as far as the historical and consolidated coastal city
- The "Landscapes of the river and torrent beds" for which mitigation measures are envisaged for the hydraulic risk and the more comprehensive system of integrated risks through operations of gradual renaturation of natural streams and river beds and reconversion as tree-lined avenues for buried waterways in urban contexts together with the preservation and enhancement of the Ganzirri and Faro lake landscape and of historical hydraulic devices
- The "Rural-Landscape matrices" for which interventions are envisaged for safeguarding and valorising the agrarian landscapes on the slopes along the valleys of the rivers, the agricultural landscapes of the valley bottom characterised mainly by citrus groves and coastal agricultural landscapes in the sub-plain areas, together with the incentivisation of a process of consolidation of the agrarian terracing in use and of restoring abandoned ones which strongly characterise the historical agricultural landscape

[14] See Document "P2.2 Settlement System and Services. Project Guidelines" of the Preliminary outline of the new City Plan of Messina. P2.2 Sistema Insediativo e dei servizi. Linee Guida di progetto" dello Schema di Massima del nuovo PRG di Messina. Available at Comune di Messina. https://www.comunemessina.gov.it/area-pol-territorio/schema-di-massima-prg-2018/

[15] See the report of the preliminary outline of the new City Plan, which identifies a strategic framework articulated in priority objectives and strategic features to which some *visions* of cities correspond: (1) "City-mosaic of excellent landscapes"; (2) "Resilient and anti-fragile city"; (3) "Polycentric, regenerated and habitable city"; (4) "City-intersection, interconnected and accessible"; (5) "Attractive, welcoming and creative city"; (1) "Città-mosaico di paesaggi eccellenti"; (2) "Città resiliente e anti-fragile"; (3) "Città policentrica, rigenerata e abitabile"; (4) "Città-snodo, interconnessa e accessibile"; (5) "Città attrattiva, accogliente e creativa". Available at Comune di Messina. https://www.comunemessina.gov.it/area-pol-territorio/schema-di-massima-prg-2018/

- The "Linear coastal landscapes" for which erosion mitigation, beach nourishment and redevelopment operations are envisaged, together with the strengthening of retrodunal vegetation
- The "Urban ecological constellation" for which, on an urban scale, regeneration interventions, valorisation and implementation of the existing system and the potential of micro and macro green pores are envisaged to rethink the relationship between cities and open spaces, incentivising sustainable management of soils and water, implementing the provision of urban standards through a new offer of multi-functional and leisure spaces, and, on the regional scale, actions for ensuring that safety standards are respected, reclamation and recycling also as large parks for quarries and disused tips and activities, if they are to be abandoned

14.6 The GBIs in Support of the Strategic Flagship Projects and Intervention Priorities

This dense network of GBIs, defined starting from the existing ones, moves within a dual-scale dimension, regional and local, to contrast the conditions of fragility and stress by identifying a set of multi-scalar and inter-systemic actions to reduce the exposure and vulnerability, maximise biodiversity and the production of ecosystem services, also in urban areas, thus contributing to ensuring that safety standards are respected in the areas at risk and regeneration of the territory (Center of Neighborhood Technology CNT 2010). Indeed, although they have a systemic approach, they take shape through places, resources and practices, representing an open network of multifunctional (Fabos and Ahern 1995) and multi-scalar relationships that become central in the rethinking of a new vision of the city for experimental planning of places of landscape and ecological quality, traversed by inclusive social practices, innovative economies and collaborative public–private processes, starting from the identification of the three Strategic Flagship Projects (SFP):

- "The great metropolitan park of the Peloritani in the city itself and the 'comb' of the rivers" as an opportunity for the city to have a park that is adequate to the real urban dimension and the environmental regeneration of the present fabrics infiltrated by the GBIs inside which there are rivers desaturated by building pressure and made safe, valorised and certified in terms of equipment and services, in order to reconstruct the balance of relationships between the city, the waters and the quality of the soils and their uses (Fig. 14.4).
- "The sequence of coastal landscapes and excellences on the two seas" that, along the large blue infrastructure of the coastline, intercepts a sequence of places and landscapes to be made safe and valorised in their environmental and functional diversity, from Giampilieri and the Port of Tremestieri, to the former industrial areas that have been regenerated and the historic Port of Falcata up to Capo Peloro and beyond, towards Mortelle Tono and Villafranca. The development of

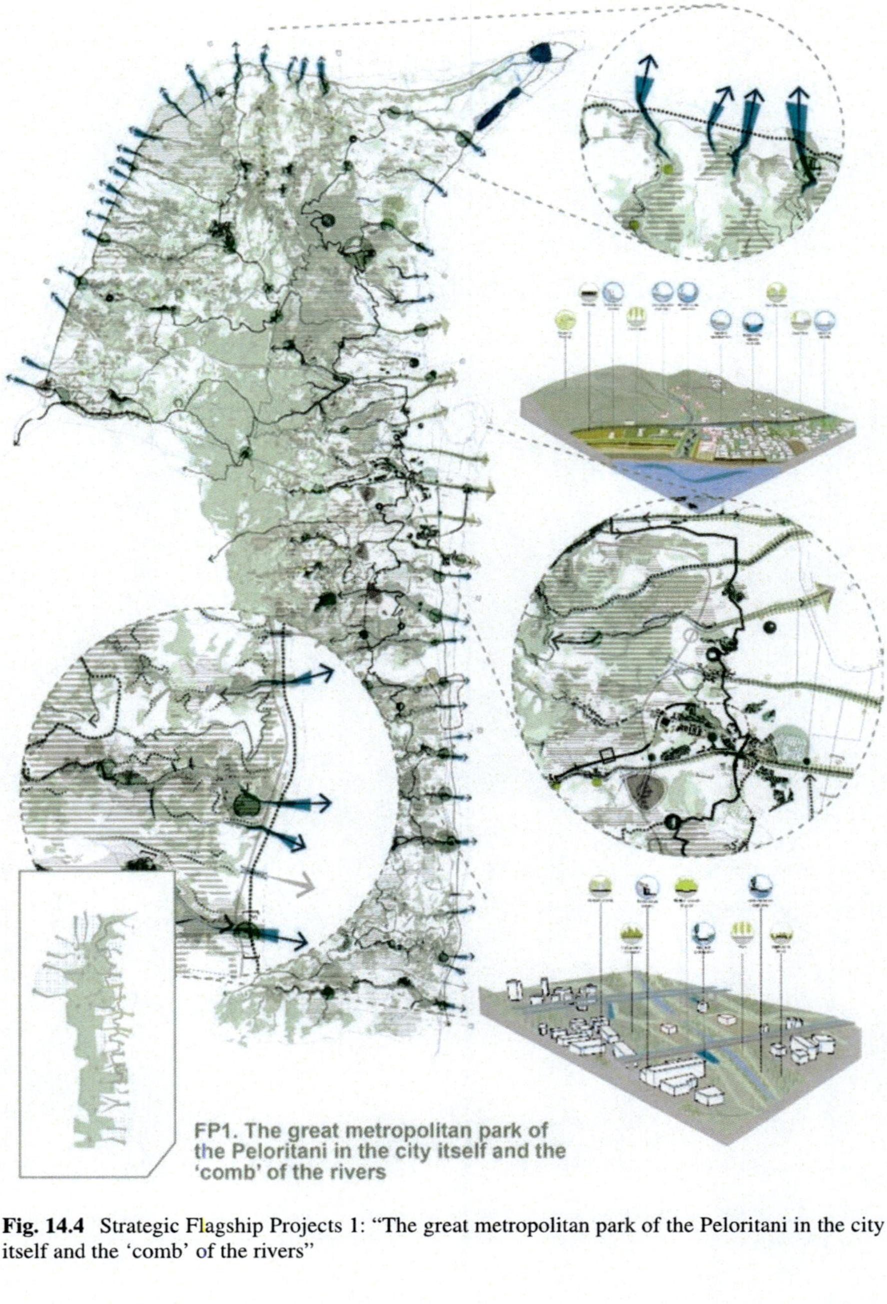

Fig. 14.4 Strategic Flagship Projects 1: "The great metropolitan park of the Peloritani in the city itself and the 'comb' of the rivers"

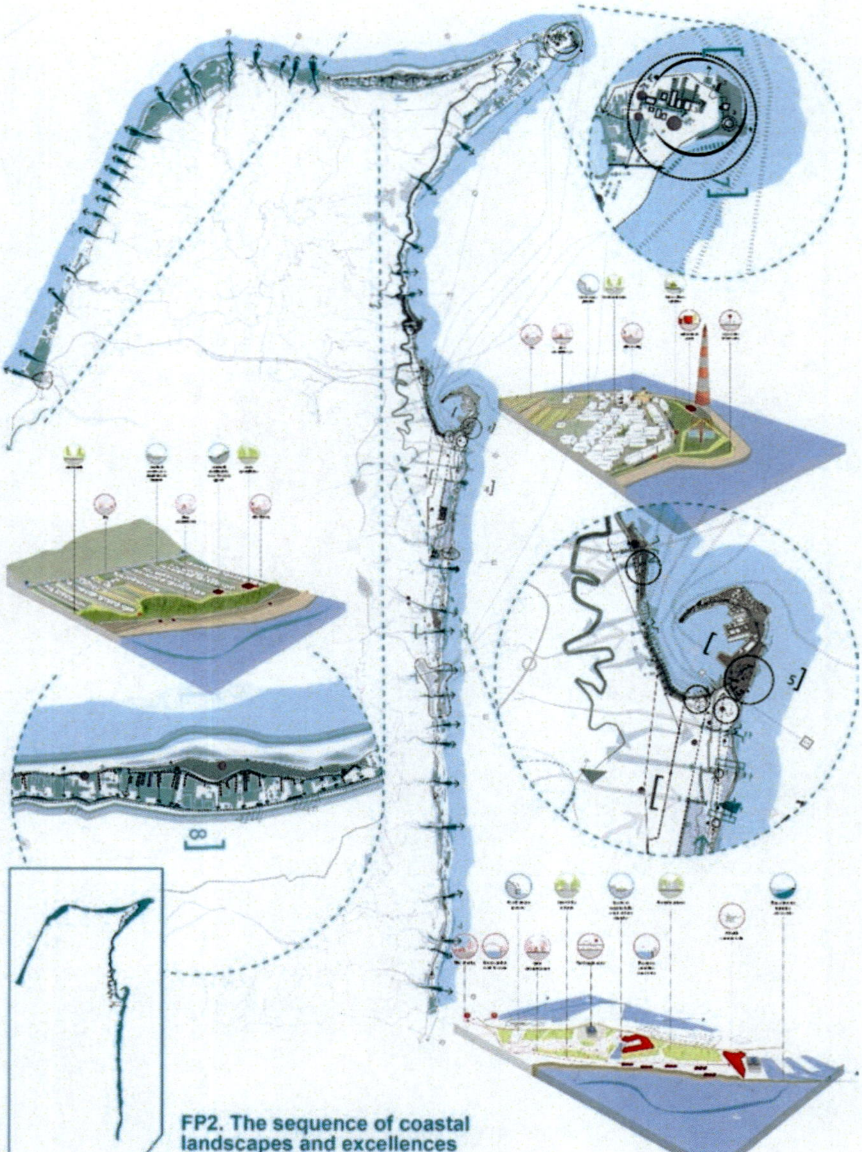

Fig. 14.5 Strategic Flagship Projects 2: "The sequence of coastal landscapes and excellences on the two seas"

slow cycling and pedestrian mobility and continuous public transport also favours the widespread accessibility to some urban and local scale centralities to be characterised and promoted (Fig. 14.5).

- "The disused railway as a greenway for urban regeneration" along the Messina-Palermo railway line, which cuts across the municipal territory from coast to coast, from the Ionian Sea to the Tyrrhenian Sea. Its recycling and re-functionalisation as a greenway, and therefore as a linear public space, activates a series of urban regeneration interventions of the settlements with the greatest urban and social degradation starting from the initiatives already being undertaken with national and international funds by the "CapaCity" project and POC Metro.[16] At the same time, the Linear Park defines a strategic axis in the broader system of slow mobility, connecting it to the coastal circuit to close an urban cycle and pedestrian *ring road* as the backbone of a minute network of connections, footpaths and pedestrian paths that innervate the city from the sea to the mountains (Fig. 14.6).

Thus, in the parts of the city involved in the SFPs, there are the same places and areas of greater malleability and propensity to change traversed by the GBIs, and these represent the priority design and implementation opportunities because they are not only able to accept the main fields of convergence of the action public but also give shape to the main issues, be they central or cross cutting, of urban regeneration identified within the framework of strategic objectives and for which the efforts and resources of the various public administrations and private subjects involved in the implementation of the City Plan will be concentrated.

It is therefore clear that GBIs and SFPs inform each other through a circular dynamic in which the safeguarding, protection and requalification of the natural and anthropogenic landscapes that characterise and qualify the city and the territory of Messina are expressed through the strengthening of the great territorial ecological connections and matrices (Fabos and Ryan 2004). At the same time, the consolidation and strengthening of the widespread system of urban permeability, the protection of biodiversity and eco-sustainability of farming systems and the quality of the landscapes aim to produce not only environmental but also economic effects, due to the tourist vocation, of which the variety of landscapes in Messina is an essential component.

Note The reflections contained in this essay refer to a wider work developed during the drafting process of the Preliminary Plan of the City Plan of Messina (approved by the City Council on April 26, 2018). A selection of drawings and the General Report can be examined on the website of the Municipality of Messina (http://www.comunemessina.gov.it/area-pol-territorio/schema-di-massima-prg-2018/).

The project team of the Municipality of Messina: Ing. Antonio Cardia, Director of the Department of Territorial Policies; Dr. Placido Accolla, Urban Plan technical

[16] http://www.ponmetro.it/home/programma/cosa-e/programma-azione-coesione-complementare/

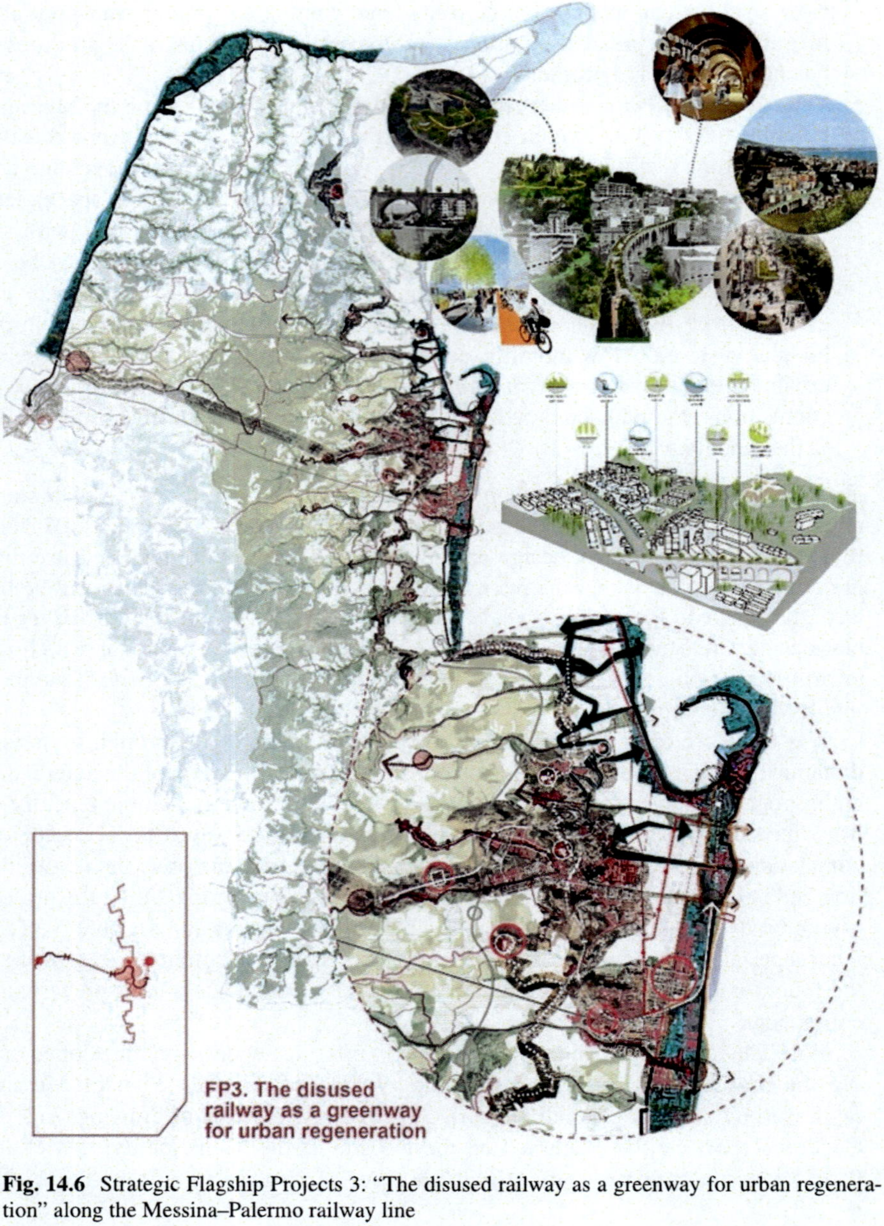

Fig. 14.6 Strategic Flagship Projects 3: "The disused railway as a greenway for urban regeneration" along the Messina–Palermo railway line

and administrative Manager. General Consultant of the City Plan: Prof. Arch. Carlo Gasparrini.

The project team supporting the General Consultant Prof. Arch. Carlo Gasparrini: arch. Daniele Caruso, arch. Stefania D'Alterio, arch. Cinzia Panneri, arch. Roberto Riccio, arch. Valeria Sassanelli, arch. Anna Terracciano; interns of the University of Naples "Federico II," Dr. Giovanna Ferramosca, Dr. Rocco Orefice, Dr. Carmen Prisco, Dr. Stefano Spera.

References

Beck U (2013) La società del rischio. Verso una seconda modernità, Quality paperbacks, Carocci

Benedict MA, McMahon E T (2006) Green infrastructure: linking landscapes and communities in landscape ecology, vol 22, issue 5. Island Press, Washington, DC, pp 797–798

Center of Neighborhood Technology CNT (2010) The value of green infrastructure: a guide to recognizing its economic, environmental and social benefits, Chicago, Illinois, Center for Neighborhood Technology and American Rivers. Available at CNT. https://www.cnt.org/publications/the-value-of-green-infrastructure-a-guide-to-recognizing-its-economic-environmental-and

Comune di Messina (2018) Piano regolatore generale. Schema di Massima. Relazione. Available at Comune di Messina. https://www.comunemessina.gov.it/area-pol-territorio/schema-di-massima-prg-2018/

Council of Europe (2010) Convenzione europea del paesaggio. Available at COE. https://rm.coe.int/CoERMPublicCommonSearchServices/DisplayDCTMContent?documentId=0900001680080621. Accessed 20 Oct 2010

De Cola S, Gasparrini C (2017) Il Piano d'Azione "Ricostruzione del quadro informativo sull'esposizione ai rischi naturali del patrimonio abitativo del Comune di Messina". In: Rapporto sulla Promozione della sicurezza dai Rischi naturali del Patrimonio abitativo, Struttura di Missione Casa Italia, Presidenza del Consiglio dei Ministri. Available at Dipartimento Casa Italia. http://www.casaitalia.governo.it/it/approfondimenti/rapporto-sulla-promozione-della-sicurezza/

ENEA (2013) Valutazione della pericolosità da frana nel territorio del Comune di Messina. Available at ENEA. http://openarchive.enea.it/handle/10840/4771?show=full

European Commission (2018) Report from the Commission to the European Parliament and the Council on the Implementation of the EU Strategy on Adaptation to Climate Change. Brussels, 12.11.2018 COM(2018) 738 final. Retrieved from https://www.eea.europa.eu/

European Union (2009) Libro Bianco. L'adattamento ai cambiamenti climatici: verso un quadro d'azione europeo. Available at EUR-Lex. https://eur-lex.europa.eu/LexUriServ/LexUriServ.do?uri=COM:2009:0147:FIN:it:PDF. Accessed 31 Apr 2009

European Union (2010) Biodiversity strategy for 2020. Available at EEA. https://www.eea.europa.eu/publications/eu-2010-biodiversity-baseline/. Accessed 31 May 2010

Fabos J, Ahern J (1995) Greenways: the beginning of an international movement. Elsevier, Amsterdam

Fabos JG, Ryan RL (2004) Editorial: international greenway planning: an introduction. Landsc Urban Plan 68(2–3):143–146

Forman RTT, Godron M (1986) Landscape ecology. Wiley

Gasparrini C (2017a) The resilient metamorphosis of the cities". In: The Plan Journal "Resilient edges" n. 2

Gasparrini C, Terracciano A (eds) (2017b) Drosscity, ListLAB, Trento-Barcelona

Gasparrini C (2019) Infrastrutture verdi e blu. Una priorità nazionale per la pianificazione urbanistica e la coesione territoriale nei prossimi anni. In: Urbanistica Informazioni n. 282
LOTUS (1999) Tutto è paesaggio, n. 101
McHarg IL (1997) Ecology and design. In: Thompson GF, Steiner FR (eds) Ecological design and planning. Wiley, New York, pp 321–332
Mell IC (2008) Green infrastructure: concepts and planning. FORUM: Int J Postgraduate Stud Archit Plann Landsc 8(1):69–80
Mell IC (2015) Green infrastructure planning: policy and objectives. In: Sinnett D, Smith N, Burgess S (eds) Handbook on green infrastructure: planning, design and implementation. Edward Elgar Publishing Limited, Cheltenham. Available at E-Elgar. https://www.e-elgar.com/shop/eep/preview/book/isbn/9781783474004/
Mostafavi M, Doherty G (2010) Ecological urbanism. Lars Muller Publishers
Sinnett D, Smith N, Burgess S (2015) Handbook on green infrastructure: planning, design and implementation. Edward Elgar Publishing Limited, Cheltenham. Available at E-Elgar. https://www.e-elgar.com/shop/eep/preview/book/isbn/9781783474004/
SIT Regione Sicilia, CUAS (2013) Carta dell'uso del suolo agricolo, Retrieved from http://www.sitr.regione.sicilia.it/dawnload-cartografia
United Nations (1987) Brundtland Report. Report of World Commission on Environment and Development "Our common future". Available at ARE. https://www.are.admin.ch/are/it/home/sviluppo-sostenibile/cooperazione-internazionale/l_agenda-2030-per-uno-sviluppo-sostenibile/onu-_-le-pietre-miliari-dello-sviluppo-sostenibile/1987%2D%2Drapporto-brundtland.html
United Nations (1992) Agenda 21. United Nations Conference on Environment & Development Rio de Janerio, Brazil, 3 to 14 June 1992. Available at UN Sustainable Development Goals. https://sustainabledevelopment.un.org/outcomedocuments/agenda21
United Nations (2012) Il futuro che vogliamo. Atti della conferenza Rio +20 United Nations Conference on Sustainable Development. Available at ARE. https://www.are.admin.ch/are/it/home/sviluppo-sostenibile/cooperazione-internazionale/l_agenda-2030-per-uno-sviluppo-sostenibile/onu-_-le-pietre-miliari-dello-sviluppo-sostenibile/2012%2D%2Dconferenza-sullo-sviluppo-sostenibile%2D%2Drio-20-.html
United Nations (2015) Trasformare il nostro mondo: l'Agenda 2030 per lo Sviluppo Sostenibile. Available at UNRIC. https://www.unric.org/it/images/Agenda_2030_ITA.pdf. Accessed 21 Oct 2015
United Nations (2019) UN environment's sixth global environment outlook. Available at UN Environment. https://www.unenvironment.org/resources/global-environment-outlook-6
United Nations Environment Programme (UNEP) (2012) 5th Global Environment Outlook (GEO-5) "Environment for the future e want", Malta, Retrieved from https://www.unenvironment.org/resources/global-environment-outlook-5
Waldheim C (2006) Landscape urbanism reader. Princeton Architectural Press, New York
Wolman A (1965) The metabolism of cities, Scientific American, n. 213

Chapter 15
Green Texture: Nature and Reuse in the Prato Operative Plan Legislation

Francesco Caporaso, Pamela Bracciotti, and Antonella Perretta

Abstract Prato is a city rich with history and environmental resources, characterized by polycentric urban development, where nature invades the residential area and crowns it: a large agricultural plain to the south and mountains to the north. The city has undergone significant demographic growth; it is a working city and place for artistic and urban experimentation, full of resources and contradictions, like many medium-sized cities. The "Operative Plan", a new urban planning tool, combines the structures Prato inherited from its industrial history in the textile sector with new infrastructural projects. Ongoing projects are inserted into urban areas and green spaces, while taking into account the need to address climate change, limiting the use of soil, recovering and reusing existing buildings and maintaining their identities through a regulatory framework designed to give the city a new nature, creating a new urban fabric.

Keywords Polycentric · Inherited · System of canals · Circular economy · Healthy city · Green benefits · Urban forestation · Urban demineralization · Urban regeneration

15.1 The City of Prato

15.1.1 The City

> *"Prato is where the history of Italy and Europe comes to rest: every rag ends in Prato [....] For years the people of Prato have woven, knitted, carded the rags from Marengo, from Austerlitz, from Waterloo, the flags of the Grande Armée, the uniforms of Murat, the golden tailcoats of the Holy Alliance [....]. In Prato, where everything comes to rest: glory, honor,*

F. Caporaso (✉) · P. Bracciotti · A. Perretta
Servizio Urbanistica – Ufficio di Piano, Comune di Prato, Italy
e-mail: f.caporaso@comune.prato.it

A. Arcidiacono, S. Ronchi (eds.), *Ecosystem Services and Green Infrastructure*, Cities and Nature, https://doi.org/10.1007/978-3-030-54345-7_15

pity, pride, the vanity of the world [....] this is what to keep in mind: Prato is a blue collar city, all workers, the only one in Italy, from head to foot" (Malaparte 1956).

With nearly 200,000 inhabitants, Prato is the third largest city in central Italy after Rome and Florence, thanks to the immigrants that arrived first from the countryside, then from southern Italy, and finally from abroad. The city underwent a demographic boom that exceeded the national average, making it a lively, dynamic place. More than 130 foreign ethnic groups make up 18% of the resident population, with the largest Chinese community in Europe standing out, all making Prato at the end of 2004 the Italian province with the highest percentage of foreign population among residents.

Prato is known not only as a multiethnic metropolis but also as a multicultural metropolis, thanks to the presence of three large universities: the PIN Polo University, Monash University (Australia), and New Haven University (USA). Prato is a contemporary city where the presence of the Luigi Pecci Contemporary Art Museum stands as the first of its kind founded in Italy (1988). With 3500 companies in the textiles district and 4000 companies in the fast fashion district, Prato stands among the leaders in Europe and is a point of reference for large-scale distribution, in unison with the ICT sector, with approximately 700 companies and young start-ups. The territory of Prato is characterized by large open spaces intersected by "green belts." These are large green axes crossing through the city, connecting the agricultural plain to the hills of Calvana and Monteferrato – and "farming squares" – green spaces in various sizes that encroach into the center of the city – that together form a large internal crown that reflects the outlying Florence–Prato–Pistoia Plain. This system is accompanied by a lasting "polycentric" organization of the city: a historic center marked by city walls, where expansion never took hold around the crown of nuclei and historical neighborhoods acting as satellites. Already a preferred place of the Etruscans,[1] situated in the Florentine plain between Florence and Pistoia, Prato is characterized by its polycentric structure: an urban center enclosed within the walls erected in the 1300s,[2] with a castle built by Frederick II,[3] surrounded by several smaller centers around the churches or workshops. Their presence was also favored by the driving force of the abundant waters flowing here, which through a system of canals flow down through channels to the entire city,

---

[1]Archeological findings seem to point to structured colonization in the Iron Age, with acknowledged Etruscan urban settlements used from the sixth century B.C. "Prato e la sua Provincia" by C. Cerretelli, ed. Giunti, 2003.

[2]The second ring of city walls built in the seventh century enclosed the *burnus* and the *castellum and divided the castrum* into four neighborhoods that formed a fortified settlement. Demographic growth required expansion of the walls in 1270 to the south, but those centuries were marked by positive periods alternating with years of epidemics and wars, with a subsequent drastic decline in the population. At the end of 1384, the city walls were completed and measured approximately 4.5 km in their perimeter.

[3]The Castle of Frederick II, built toward the end of 1230 when Prato was chosen to be the headquarters for the deputy of Frederick II and the Alberti fortress, was taken over by the construction of the castle. "Storia illustrata di Prato", F. Cardini, ed. Pacini, 2003.

almost 50 km of derivative streams[4] along which the first water mills and fulling mills were erected, followed by the proto-industrial buildings that covered the entire textile-making production cycle. Recovery of rags and then reusing wastewater from the industrial aqueduct then made Prato a city that was naturally focused on a circular economy, paradoxically before this was even considered and coined as a concept.

15.2 The Operative Plan

15.2.1 A New Vision for the City

The Operative Plan,[5] a 5-year urban planning tool, sets forth detailed provisions for how and how much it is possible to intervene in the transformation, enhancement and protection of the municipal area, starting from a general overview of social, cultural, and economic development in the city, already formulated by the Administration when shaping public policy.[6]A plan was composed of over 100 survey, descriptive, and prescriptive[7] documents, with the objective to present an accurate picture of the environmental, historical, and cultural resources, to guide conservation, development, transformation, and limit impact. The Plan has the ambitious aim of defining local sustainable developmental models from a social, cultural, and economic point of view for the city's future, taking responsibility for the specific character of Prato and making it a central aspect in terms of the ongoing European and national debate over urban planning and reusing the existing environment. These two themes are very closely connected and are the constant thread that runs through all the project decisions in the Plan, which are focused, in terms of strategic technical terms, on reducing use of soil, resilience, and quality of the urban environment, the capacity of the city to confront themes related to climate change, heat islands, and air quality.

One of the decisions of the Administration was to assign the Planning Office with drafting and implementation of the Operative Plan, an urban planning tool for the city. It is made up of people with proven experience and an intimate, comprehensive vision for development of the city, and, in this sense, who are able to check the actual state of progress and efficacy of the implemented strategies. The

[4] http://www2.comune.prato.it/ambiente/cavalciotto-e-gore/

[5] Adopted by DCC no. 71 on 06/09/2018 and approved by DCC no. 17 on 04/03/2019

[6] The public policy track, approved by DCC 89/2015, and subsequently the Procedure Initiation document, approved by DCC 86/2016, making the best use of the previous plans (Secchi Plan of 2001 and the Structural Plan of 2013, the Extraordinary Plans and municipal plans in the PAES and PUMS sector) attempted to provide answers for the changing needs of the city through a new idea of the city, impacted by numerous layers of change and subject to continuous and unrelenting future change.

[7] http://www2.comune.prato.it/piano-operativo/

foundational strategies of the Operative Plan, that is, the Manufacturing of the twenty-first century, or Reuse, the Grand Projects or the Strategic Areas, the Ecological, Environmental and Agricultural aspects, the Public Space or the Urban Policies of Welfare, all required a landscape/territorial humus. The hotbed consists of signs to be rediscovered or highlighted, transforming the environmental or ecological themes that in the Track Policy was one of the strategies for development into a common thread. The "environmental theme" from the "monthly event" in the process became the crossover and unifying theme as well as the central supporting axis of the Plan. It is an innovative vision that adheres to the philosophy of the "healthy city," in terms of the project, translating a new conception of open space and nature in the city to the urban scale, an authentic active tool for public health. The results of this extended information-gathering, participatory process produced a regulatory framework focused on the quality of the environment and the urban and rural landscape that make up the territory. The intention is to give the city a green infrastructure, evident in each single article, creating a fabric for weaving in construction and infrastructural interventions, corresponding to actions designed to increase the natural assets of the city. The Plan is designed to limit consumption of soil through the recovery and reuse of existing buildings. A specific research project was aimed at high-quality construction: the result is a one-of-a-kind photograph where some buildings from the 1500s (or even the Middle Ages), buildings from the 1700s and early 1900s, with evident agricultural roots, stand alongside industrial buildings and complexes that marked the history of productive Prato. Each of them is accompanied by an information sheet and specific regulatory document to help the professionals who will be working on them as part of a comprehensive framework built on the knowledge of the historical and economic values of the building for its owners and the collective memory of the entire city. For the more recent buildings, which are in disuse or no longer functional, it is possible to regenerate them through partial or total demolitions, rebuilding with heights and configurations that are more favorable to creating or expanding green areas, in a gradual demineralization process, where walls and roofs are slowly invaded by renewed nature. The planned new buildings are permitted based on substantial transfers to the public of unbuilt areas inside the urban territory, offering the possibility to insert sports facilities, recreation centers, and areas for free-time or public services, which simultaneously act as a protection against climate change, improving noise levels and visual impact, air quality and quality of life in general (Fig. 15.1).

15.2.2 The Environment

Urban Planning in Prato – Experimental City Since the post-war period, Prato has been a permanent experiment in urban planning; Nello Baroni Plan 1954, Leonardo Savioli Plan 1954–1956, Plinio Marconi Plan 1961–1964, Sozzi-Somigli 1975–2001 Plan, and Bernardo Secchi Plan 1993–2019, the authors of the Regulatory Frameworks, for the excellent Urban Planning Construction, Structure

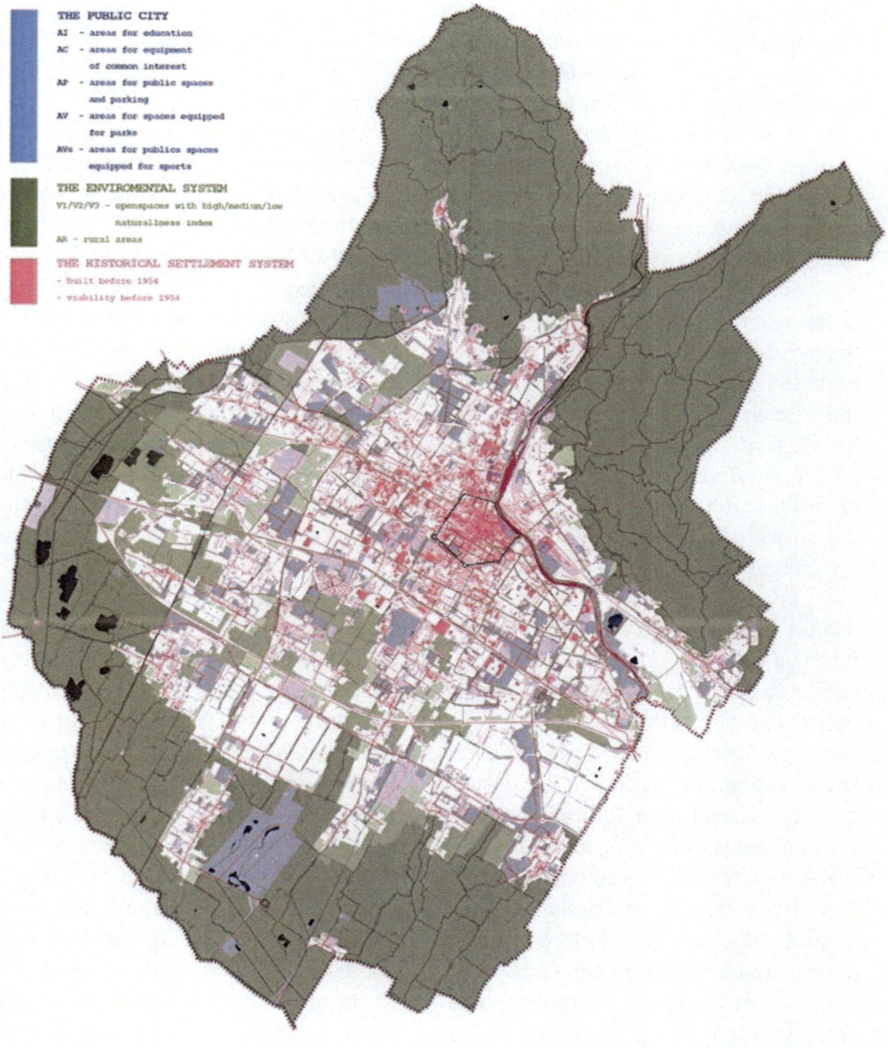

Fig. 15.1 Prato Municipality, Map of the public city, Exposition Panels RUN 2019 INU – extracts, Operative Plan (Prato 2019a, b, c)

and Regulations, joined by the experience of the "Radical" Florentine group and interests for the factory city.[8] The Operative Plan followed in these footsteps, cre-

[8] For more information, refer to P.M. Vannucchi, Le fasi della pianificazione urbanistica a Prato, Poggibonsi, 2008 and http://www.comune.prato.it/servizicomunali/prg/nuovops/pianificazione/02/

ated "in-house" by the Municipal planning group with the help of important external consulting for specific topics, in particular for the Urban Forestation and Green Benefit strategy,[9] by Stefano Boeri Architetti, Prof. Stefano Mancuso – PNAT and for urban equalization, Prof. Stefano Stanghellini.

"Green Benefits" The survey of live trees in public spaces in Prato, counted 29,151 trees in 2015, which was the basis for the benefit study on live trees in the city. The available data, thanks to the use of dedicated software like i-Tree ECO, which analyzes the species, structure and age, biomass and leaf index, combined with data on the local climate and atmospheric pollution, made it possible to estimate the benefits of trees and assess their actual value in the future. It was also possible to estimate the percentage of green coverage areas in the city thanks to aerial imagery. Coverage included private trees and areas outside of the survey area. The model demonstrates the percentage of land coverage, at 65.4% plants and trees (composed of 23% trees, 38.5% farming areas and pastures, 3.9% shrubs), with the remainder composed of 13.8% buildings and 20.8% asphalt. These data are essential for guiding planning choices, demonstrating that the most conspicuous consumption of land is asphalt, not only roads, but also large parking areas without any trees, which are responsible for increased local temperatures.

Action Plan for Urban Forestation An interesting tool for guiding the choices made in the Operative Plan, based on the concept of urban and semiurban forestation, a management system for metropolitan forests. It is useful for guaranteeing optimum contribution to physical, social, and economic well-being in urban societies, to be implemented through an integrated, interdisciplinary, participatory, and strategic planning approach. Forestation associated with correct planning can make the city resilient. The Action Plan for Urban Forestation has the objective to increase forested areas in the city, in particular in the areas with high rates of urbanization, to restore spaces and corridors for life through urban re-naturalization processes; a large forest that unites existing and newly planned parks, farming areas, and private natural areas into a single system; a planned system including 190,000 trees, approximately one for each resident, to be planted in concentrations, sparsely or in rows, accompanied by shrubbery and grass; an effective tool for environmental compensation, which is implemented in the regulatory framework, figuring as a new program. This is a project that cannot rely on specific potential financing and therefore requires collaboration between the public and private spheres for its implementation, starting from the diffusion of best practices. The different types of forestation include semiurban forests, public parks, with surface areas exceeding 5000 m^2, small parks, with surface areas less than 5000 m^2, and green buildings (green roofs or facades, or use of low-emission materials). The Action Plan lays out six implementation strategies:

---

[9] http://pubblicazioneatti.comune.prato.it/pubblicazionebinj/servlet/RicercaDelibere?ente=001&dadata=14.03.2019&adata=14.03.2019&danumero=17&anumero=17&tipoAtto=CC&anno=2019&paroleTesto=&tipoRicercaTesto=&sort=&delXPag=10&parole=&tipoRicerca=&ordina=

1. The River and Gore park – The Bisenzio park and Gore waterway network
2. Green planting to mitigate infrastructures – rows of trees and mitigating areas in the road and railway networks
3. Widespread green planting – small intersecting green areas and microregulation in the urban environment
4. Semiurban farming belts and large parks – Semiurban farming belts and new synergies among urban, farming, and natural areas
5. Urban demineralization – reduction of impermeable areas and urban decontamination through demineralization and naturalization
6. Local agricultural park – enhancement of local productions and supply chains

The six strategies can count on case studies as well as on specific schedules for implementation of forestation policies: actions, interventions on buildings, main tree species and shrub species, the pilot project in the San Paolo park, using all the tools made available according to a set of obtainable objectives. An atlas was also drafted to local scale, which reproduces the effects of the planned urban forestation interventions in the city (Fig. 15.2).

15.2.3 The Contemporary Image of the City

The Prato Municipality, in designing its intense urban planning policies, in alignment with international objectives, has advanced projects aimed at improving the urban resilience of the city, and more in general at increasing the environmental character of the policies, to improve the quality of life of its citizens. The Operative Plan is the strategic planning document for the medium-long term, which summarizes all the decisions made in relation to the city, launching it toward new future scenarios. The social, economic, and cultural objectives are aligned with the priority objective to construct a sustainable city, maximum savings of all available local resources, and most importantly those resources that cannot be reproduced or can only be reproduced with extensive time and costs. Those resources constituted by residual natural elements that persist in their essence and interpretation are considered to be essential. It is therefore necessary to reconnect, reuse, recognize, recover, and qualify (Di Bello 2015). The need is not so much to quell expansion, since growth in Prato hovers around 1200 units annually, but to guide, facilitate, and organize transformational strategies, where green areas are not considered an element of contrast to the built-up city, but instead an element for a new landscape composition. The unconstructed spaces, public spaces, become places for building future landscapes. Cultivated areas, extended forests, the banks of rivers are the places for this new design, for the construction of new environments (Di Bello 2015). The Plan project is therefore a "territorial re-composition" expressed through two main elements: public space and the landscape. The project aims at hemming together spaces, enhancing the territory and existing ecological connections, from a territorial scale to the details of public spaces, connecting places and territories through

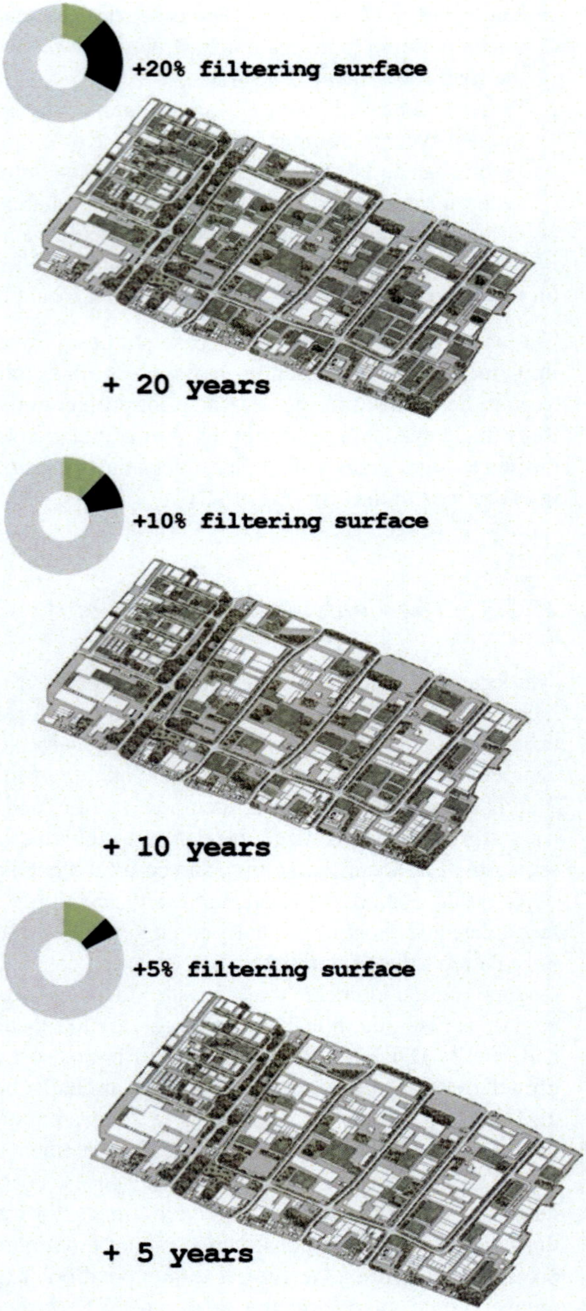

Fig. 15.2 Prato Municipality, "Demineralization interventions in the area Macrolotto 1" Exposition Panels RUN 2019 INU – extracts, Operative Plan, (Prato 2019a, b, c)

the elements they are composed of. The city and the natural surrounding environment are therefore connected through a variety of public areas and spaces, part of a synergy where one increases the ecological role and value of the other. A continuity, a reticular dimension pervades the territories, unites them, and guides their development: flexible, articulated, and differentiated, in contrast to the fragmentation and lack of relations that are so evident today. Water, together with trees, are primary elements for regulating temperature and ensuring ecological continuity, fundamental for defining strategies that can combat climate change. The plan therefore moves in the direction of more extended use, with the exception of usage limits in place to save water, and the presence of pollutants in waterways and aquifers.

The relationship of Prato with water is a story of deep connections between the natural element and humans, which over the course of centuries witnessed the anthropization of the land linked closely to the creation of innovative strategies for managing water. The Gore canals system is part of this human–water dynamic, inherited from the history of Prato, one of a kind in Europe, which the Operative Plan valorizes in its many dimensions. Among them, a territorial system that interconnects open spaces, nature, and buildings; a historical landmark; an ecological and physical connection between parts of the city; and a potential source for the production of renewable energy. This city/water bond is also represented by the persistence of innovative investments, which over the course of 40 years have resulted in one of the most important systems for reusing urban wastewater (civil and industrial), with the realization of the industrial aqueduct that reuses water. This infrastructure is a fundamental part of the textiles production industry in Prato and opens up new scenarios and potential for future human/water possibilities, linked to new city management models based on circular economy principles. The Operative Plan, thanks to acquired knowledge, takes these strategies to heart and transforms them into specific actions and interventions. Taller buildings are isolated as new urban landmarks that are part of a new urban skyline, contrasting the heat islands and letting airflow, cooling temperatures. The progressive reduction of some denser parts of the city in favor of usability and permeability of the urban fabric, accompanied by the creation of urban empty spaces for the purpose of forestation, are handled with awareness, to prevent distortion of recognizable historical/residential characteristics. A new skyline runs along the axis of the Declassata, where taller buildings are situated alongside smaller ones with roof gardens, parks, and tree-lined parking areas, in an orthogonal scheme along the roadway axis that follows the footsteps of centuriazione. The plan inspires the dream of a "green" bridge that goes over the Viale da Vinci to connect the Parco delle Fonti with the former Banci area. A north/south connection that runs along the linear Parco del Soccorso, moving a complex that today stands in abandon closer to the city. Other parks are also feasible, San Paolo, Ciliani, and Cafaggio, which together with the "semiurban farming belts," the riverbed park, the agricultural park, and network of green parks connecting them, configure a new open view of the city.

The Transformation of Residential Areas Planning the transformations in the Operative Plan has two main goals: The first is relative to urban regeneration in abandoned productive areas. The second is the creation of new areas of public interest that increase open spaces, social gathering places, and areas for cultural events in the city. Through the transformational areas, it will be possible to create four large new urban parks: the Ciliani park, the Fonti park, the Cafaggio park, and the linear San Paolo park, for the purpose of rendering large-scale green areas open and usable to the public, all situated in densely populated areas and located in strategic points in the city, in an attempt to preserve these areas left free of expansion, often still dedicated to farming, making them available to the greater public. The Operative Plan defines the rules for managing the existing residential areas, to guide requalification actions on the residential framework through interventions to modify the morphological-functional aspect of the urban fabric. In general, the Operative Plan allows renovations and replacement construction on recently constructed buildings, to regenerate the dense areas with less volume and free up spaces for public use and green areas. To improve urban resilience, the environmental context of these areas of the city, which constitute the main "heat islands", a maximum coverage ratio of 40% was established; the buildings must be constructed with high energy performance, with a maximum height of 35 m (to liberate space on the ground and recover it in height). However, the innovative standard lies in the regulations for addressing the open areas: these must have a minimum of 30% permeable surface and at least half of this area must be planted with trees, to achieve the objective set forth in the Plan to improve the environment and livability in the city and industrial areas through urban forestation.

The Promotion of Territorial Quality If the themes of the environment, climate mitigation, psychophysical well-being, and acoustic-aesthetic improvement are the basis for the strategic and regulatory choices in the Operative Plan, unconstructed space is the expression. Whether public, private, privately owned for public use, or buffer areas, space is viewed according to overall quality: territorial quality where everyone can and must contribute. Rising median temperatures are exacerbated by the existence of impermeable structures due to the presence of buildings, the roads network, and parking areas. It is therefore necessary to recover balance among spaces, citizens, and nature, initiating a virtuous cycle made of carefully planned projects, searching for materials with high reflective quality and heat emissivity, environmental compensation provided by a higher percentage of ground permeability and forestation, and promotion of attentive, mindful use of water.

Design of Public Space Open spaces and buildings define the design of the city, in particular the equipment and public areas that are part of realizing urban planning standards. For example, parking lots have regulatory standards with the objective of contributing to the environmental character of the city with "green parking lots"; the project requires significant effort in this direction: from a minimum of 25 m^2 of parking space and space for maneuvers per parking spot, the plan increases this value to 40 m^2. The significant reduction in parking spots for the comprehensive available surface area is countered by significant natural spaces for livability and good health. Parking lots are transformed into potential gardens, where once trees reach maturity (not before 15 years after planting), 75% of the area will be under shade. Not only flowerbeds separating lanes of traffic, but entire medians with trees are planned for every 5 car stalls or neighborhood, accompanied by the use of materials with a high reflective value and thermal emissivity, to guarantee not only permeability, but also a reduction in heat islands. Green is the true protagonist of the new vision of the city and is part of every intervention, even more so when concerning parks, outdoor sports fields, and vegetable gardens. Currently, the city has a discreet portion of natural green areas, consisting of large and small parks, some of them historical, others along the river; the Operative Plan adds an ample portion of green areas to the city, which scatter across the territory to impact areas that would be lacking natural spaces if not for the Plan. Each of these large parks can represent an important environmental reserve for biodiversity and human health. The equipped parks and green areas, each with a different size and each arranged according to its own unique character, must have a significant portion of vegetation and permeable ground, not less than 80% of the entire surface area. The implemented standards have the objective of equipping the parks and green areas with sports/play equipment, walking paths, and service areas, with kiosks or similar to ensure pleasant use of the space, creating a territorial ecological network to improve environmental comfort and human health, to constitute an effective connection between existing areas and new planned spaces, both public and private. Design criteria that follow the objective of protection from sources of pollution are also part of this vision, paying careful attention to the selection of suitable plant species that are compatible with the areas and maintenance requirements.

Green Networks and Connectivity From transportation infrastructures on rails and roads to slow mobility, the topic of connection has gradually extended in its core concept, defining a network of urban forestation, called green connectivity. It is composed of a linear section of forestation planned in areas with high urbanization rates, in industrial areas, and the larger road arteries. Green connectivity, thanks to its widespread nature, brings a significant contribution to completing the quality ecological network. It reduces the mineralized surface area in the wider roadways, creates a continuous or discontinuous mitigating plant barrier in infrastructures, helps to define the urban margins, fights acoustic and atmospheric pollution, acts to mitigate the climate, and improves visual perception of the urban landscape. Additional elements that contribute to the ecological quality in the territory are private open spaces, defined based on their rate of natural space, acting as strategic or complementary areas. Specific regulation pertains to the realization of green areas, sports equipment, and private parking areas, aligned with the same regulations governing public spaces. In fact, environmental quality does not distinguish property lines: as demonstrated in the Green Benefits study, the importance of a tree is in its very existence, as the basis not only for the food chain, but also an indispensable element for breathing and the only means available for reducing pollution (Fig. 15.3).

Fig. 15.3 Prato Municipality, "Prato 2030 – vision" General Report – Urban Forestation, Operative Plan (Prato 2019a, b, c)

References

Bello G (2015) Dimensione ecologica – Una ricerca sui nuovi materiali del progetto urbano, Università degli Studi di Napoli "Federico II", Dipartimento di Architettura, XXIV Ciclo di Dottorato ricerca di Urbanistica, Napoli, pp 5–15

Malaparte C (1956) Maledetti Toscani. Vallecchi Editore, Firenze, p 102, 108, 109, 113

Prato municipality (2019a) Relazione Generale, Piano Operativo, Prato. http://www2.comune.prato.it/piano-operativo/

Prato municipality (2019b) Relazione Generale- Forestazione Urbana, Piano Operativo. http://www2.comune.prato.it/piano-operativo/

Prato municipality (2019c) Pannelli Esposizione RUN 2019 INU – estratti, Piano Operativo. https://rassegna.inu.it/catalogo-espositori/

Chapter 16
Green Infrastructure and Landscape Planning in a Sustainable and Resilient Perspective

Angioletta Voghera and Benedetta Giudice

Abstract The topic of green infrastructure is becoming more and more popular within the academic world and public bodies. In the face of climate change, it represents an urgent dilemma to be developed in both urban and landscape planning practices. Green infrastructure is indeed a well-fitted stage where to combine not only environmental, ecological and landscape elements but also social and technological ones. Within the framework of sustainability and resilience, in this chapter, we attempt to outline all the relevant issues which need to be included and discussed when speaking of green infrastructure and landscape planning. In order to analyse the Italian situation, we refer to some international experiences which have developed specific national norms and policies on green infrastructure. We conclude proposing some open issues and perspectives that emerge from current international debates.

Keywords Green infrastructure · Landscape planning · Sustainability · Resilience · Social-ecological system · Design process

16.1 Framing Sustainability and Resilience for Landscape and Green Infrastructure

Sustainability and resilience are well-known buzzwords, even though the two concepts and their meaning can be easily misunderstood, confused, and interchanged (Derissen et al. 2011; Redman 2014; Xu et al. 2015; Zhang and Li 2018). In particular, on the one hand, since the 1987 Brundtland Report introducing the concept of

A. Voghera · B. Giudice (✉)
Interuniversity Department of Regional and Urban Studies and Planning, R3C - Responsible Risk Resilience Centre, Politecnico di Torino, Torino, Italy
e-mail: benedetta.giudice@polito.it

A. Arcidiacono, S. Ronchi (eds.), *Ecosystem Services and Green Infrastructure*, Cities and Nature, https://doi.org/10.1007/978-3-030-54345-7_16

sustainable development, there has been an increased interest in urban sustainability in relation to its environmental, social, and economic issues. On the other hand, the concept of resilience has emerged within different research fields and disciplines resulting in a multitude of definitions (Davoudi 2012; Folke 2016; Meerow et al. 2016; Brunetta et al. 2019). Both concepts have a strong connection between the natural and the human environment and thus put great emphasis on social–ecological systems (Folke 2006).

Over the years, sustainability and resilience have also pervaded planning and design discourses and researches as they are widely used to understand, analyse, and cope with global transformations (climatic, environmental, and energetic, etc.). The necessary characteristics of a suitable planning and design for sustainability and resilience (Voghera 2016) are the reflective capacity, the flexibility of the process, the creativity, the inclusion of stakeholders, the integration of different action scales and multiple policies, and the robustness.

In such a perspective, the tool of green infrastructure (GI) can contribute to the achievement of concepts (Voghera and Giudice 2019) and can guide planning and design processes towards more sustainable and resilient solutions (Meerow and Newell 2017; Staddon et al. 2018). In fact, the Italian National Strategy for Adaptation to Climate Change (2015) and its Action Plan (2017) attribute to landscape planning an important role for climate adaptation, recognizing the specific contribution to defining a large scale project for ecological and landscape network. This aspect is essential for the implementation of the GI project. If we take into account one of the most common definitions of GI, the one by the European Commission (2013)[1], we can notice how GI has a strategic role in the definition of a territorial and landscape scenario for sustainability and resilience. In fact, this definition integrates three key aspects (Liquete et al. 2015):

- The idea of a network composed of territorial, environmental and landscape systems
- The centrality of planning and management
- The need to integrate the concept of ecosystem services

In this sense, GI incorporates the concepts of ecological connectivity, landscape preservation and valorisation and multifunctionality of ecosystems (Lovell and Taylor 2013; Mubareka et al. 2013; Hansen and Pauleit 2014) as it can provide multiple social and ecological benefits but also ecosystem services (Lennon and Scott 2014). Thanks to this multifunctionality, GI can also be considered as a nature-based solution which can help to enhance the social-ecological quality of a territory. It is a very broad concept which is nowadays studied as green and blue infrastructure (Arcidiacono et al. 2017) and its main aim is to balance community development needs with their related natural systems (Benedict and McMahon 2006). In fact, GI includes the development of natural protected areas and recreational

[1] The European Commission identifies GI as "a strategically planned network of natural and semi-natural areas with other environmental features designed and managed to deliver a wide range of ecosystem services." (EC 2013: 3)

activities, water management, restoration of river territories, stormwater management[2] (Ahern 2011, 2013; Demuzere et al. 2014), soil preservation, ecosystem services valorisation, and urban regeneration. Each of these elements directly involves the landscape in its multifunctional role: ecological, historical, recreational, cultural, and perceptive. GI should, therefore, be considered as a landscape network (van Langevelde et al. 1998) that gathers systems of spatial, social, and value relations in an integrated and dynamic way. It requires a coherent planning action and different landscape design scales. In particular, referring to the above-mentioned characteristics of a suitable planning and design for sustainability and resilience, researchers argue that GI promotes resilience by boosting flexibility, modularisation, redundancy, decentralisation (Wilkinson 2011), robustness, and biological diversity.

When speaking of landscape planning and GI in a resilient perspective, the literature refers mainly to some wide categories of topics: the mitigation and adaptation to climate change, the promotion of well-being and sense of community and the fostering of landscape quality. In line with the European Landscape Convention indications, GI could support the landscape enhancement by promoting ecological and biodiversity functionality, connectivity to green and recreational areas, and social inclusiveness and awareness (Landscape Institute 2013).

The European Landscape Convention favours the put into practice of different sectoral policies (i.e. economic and social development, spatial planning, culture, and environment) which need to be implemented by taking into account landscape.

16.2 Landscape Planning and Design Through GI

Since its first conceptualisation in landscape ecology (Forman and Godron 1986; Burel and Baudry 1999), the role of landscape in GI discourses is central. GI indeed, as already mentioned, supports in an integrated way environmental, social, and economic quality of territory and life. Starting from this consideration, GI can be developed through a landscape approach in planning and design at all scales.

Some inspirational guidance can be derived from some international experiences, mainly European, which have introduced GI as one of the cornerstones in the development of ordinary planning tools. In particular, in this chapter, we refer to the policies developed by France and the Netherlands.

[2] In particular, researchers refer to stormwater management and GI when speaking of resilience.

16.2.1 Insights from Two European Experiences

France and the Netherlands are almost certainly the most well-known and important experiences of landscape planning and design through GI in the European context. Indeed, they have developed an institutional approach for the wide and local scale design for the GI. In France, the GI project is strictly related to a biodiversity and landscape scenario for the implementation of the European Landscape Convention (Novarina 2015). In the Netherlands, since 1989, the national planning framework identifies a national plan to develop nature that implements the polder model approach integrating the development of new protected areas, the creation of new land for anthropic uses, and the creation of an ecological network to valorise biodiversity and contrast landscape fragmentation.

France has always given great relevance to ecological and environmental factors in its planning and, in 2009,[3] has drawn a specific policy and tool on GI, the so-called *Trame Verte et Bleue* (TVB) (Fig. 16.1). TVB are defined at a national level[4] but they should be applied at different scales of planning, from the regional to the local one. In particular, Regions have to develop a *Schéma régional de coherence écologique* (SRCE – the regional plan of ecological coherence) with the aim of defining the principle stakes of TVB at a regional scale. Despite its specificity on ecological and biodiversity reinforcement, the different elements contained in these new planning tools can also help to develop a discourse on landscape planning and design. Indeed, France has never adapted its landscape policies to the European Landscape Convention indications and still refers to the tools[5] identified by the so-called "*loi paysage,*" promulgated in 1993.[6] Since the TVB acts at different scales, including the lot one, we can point out how the landscape valorisation project is strongly related to the realisation of the TVB. In this sense, TVB helps to understand the interactions between landscape and ecology. In particular, since each Region can develop its TVB in the SRCE upon its own criteria, it is relevant to outline how some Regions selected an eco-landscape approach (e.g. the Region of Rhône-Alpes[7]).

[3] In 2009 and 2010, the French Ministry of ecology, sustainable development and energy promulgated two laws – Grenelle I and Grenelle II – aimed at integrating and enhancing ecological issues into planning tools. They refer mainly to the principles of landscape ecology.

[4] In 2014, the French National State approved the "*Orientations nationales pour la préservation et la remise en bon état des continuités écologiques*" (National orientations for the preservation and maintenance of ecological continuities).

[5] They are mainly tools aimed at preserving and valorising landscape (i.e. *Directives de protection et de mise en valeur des paysages*, *Charte paysagère* for protected areas, *volet paysager* in the local plan, *Atlas de paysages*).

[6] *LOI n° 93-24 du 8 janvier 1993 sur la protection et la mise en valeur des paysages et modifiant certaines dispositions législatives en matière d'enquêtes publiques.*

[7] In the region of Rhône-Alpes, the awareness of an increasing land take and ecological fragmentation dates back to the 1990s. Since 2006 (before the approbation of the laws Grenelle), the region has then decided to develop its own ecological network.

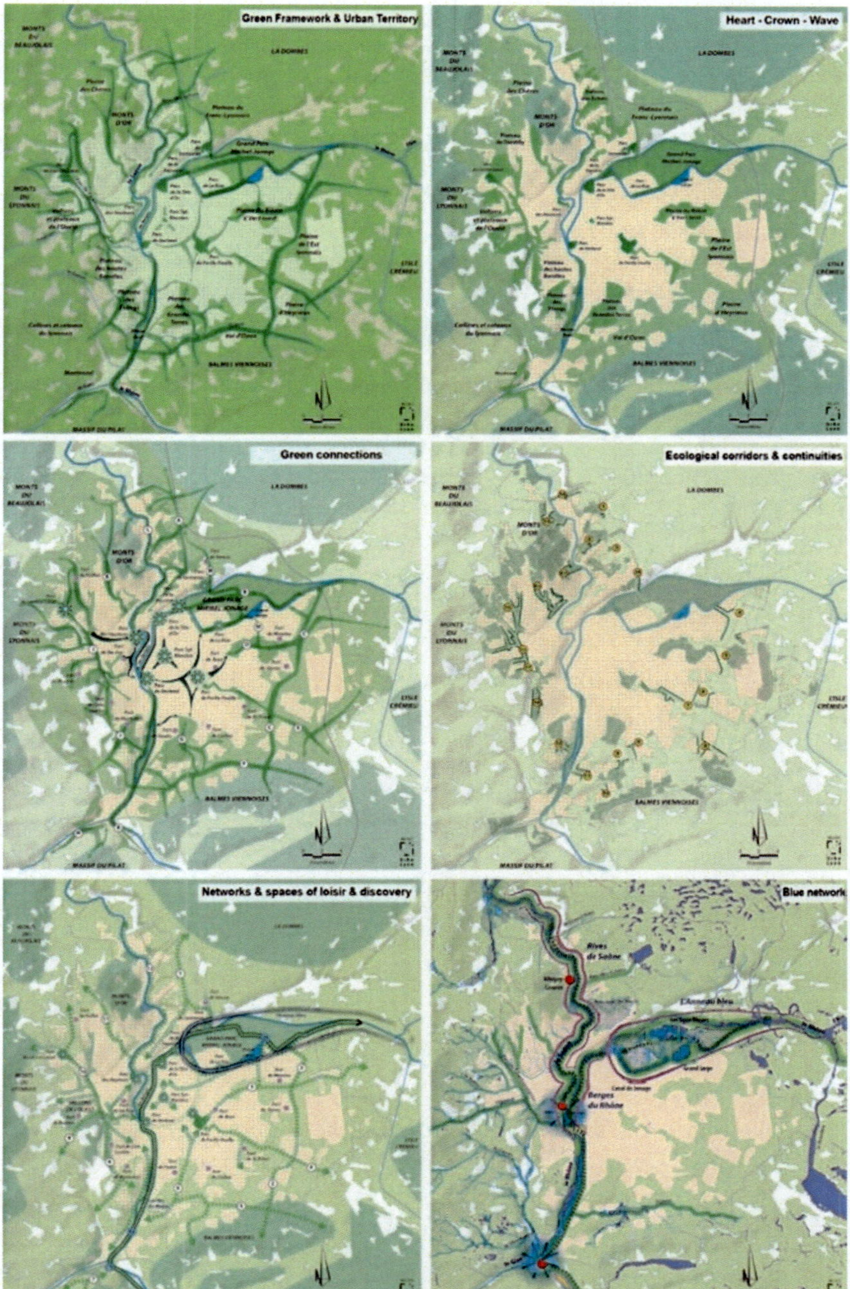

Fig. 16.1 Green and Blue infrastructure in Lyon. The design strategy. (Source: adapted from SCoT Agglomération Lyonnaise 2030)

The Netherlands have developed different sectoral policies that find integration in spatial planning at different scales (national, provincial, and local): the National Ecological Network (NEN), the Multiannual Programme Defragmentation, the National Way Forward – Government Vision 2014, the National Climate Adaptation Strategy, the Implementation Agenda Natural Capital: maintenance and sustainable use of biodiversity of 2013.

The NEN is promoted by spatial planning and protected by specific regimes in provincial regulations. The national Structure Vision on Infrastructure and Planning gives priority to this project also guaranteeing specific subsidies and funds, according to the Dutch tradition of budgeting and programming system (Faludi and van der Valk 1994)[8]. Institutional bodies in charge of realising GI projects involve different stakeholders, such as citizens, businesses and non-governmental organisations, in financing of multifunctional nature projects (van den Burg et al. 2016). In particular, projects which involve new spatial developments (affecting the water system and the natural environment) must be compensated or mitigated (Cuperus et al. 2001; Pileri 2007). Besides nature, also landscapes are protected by national and provincial spatial planning regimes. In particular, since 2017, spatial planning must also incorporate the task of 'building with nature' as a way to create nature value into urban environment.

16.2.2 The Situation of Italian Landscape Planning

In Italy, the first boost in developing green infrastructure (or, in this case, it is better to refer to ecological corridors) came from the National Strategy of sustainability and biodiversity preservation in 2010. Despite this initial endeavour, the national situation is quite stuck as Italy has not yet approved a national organic and shared project of landscape and ecological networks. While at the local scale, landscape and GI are not so evident (Voghera and Negrini 2016; Voghera and La Riccia 2018), at the regional scale, there is an increasing evidence of some of the most relevant attempts of integrating GI in planning tools. Indeed, this evidence can be traced within the context of regional landscape plans. Generally speaking, "*Italian regional landscape plans have mainly taken on a structural interpretation of landscape following a design approach*" (Voghera and Giudice 2019: 12) by assuming as the main object of preservation both the ecological value and the ecosystem services one. Following this assumption, and in line with the European Landscape Convention and the Italian Code of Landscape, some Italian regions have approved their

[8] Groene Ruimte policy in 2015 describes available subsidies for nature and landscape management: for farmers, it is estimated a budget of 30 million euro from the provinces with co-financing from the European Union of 30 million; a national budget of 350 million euro per year and an added provincial budget of around 65 million euro per year (https://www.groeneruimte.nl/index.php).

landscape plan: Puglia (2015), Tuscany (2015), Piedmont (2017), Lombardy (2017), Friuli Venezia Giulia (2018) and partly Sardinia.

In some of these plans, GI is identified as a structural component of the landscape plan (e.g. Lombardy and Puglia).

For instance, the plan developed by the Puglia region interprets the territory as heritage which integrates both material and immaterial elements; in this sense, it comprehends the social and cultural sphere and the capacity of different stakeholders to get involved. The plan has been developed through a structural approach to landscape knowledge and a strategic vision which aimed at reconnecting landscape valorisation and territorial planning at different scales. In particular, the plan identified five territorial projects for the regional landscape (e.g. the regional ecological network and the *patto città-campagna*) and 13 integrated projects of landscape, which have made possible the early realisation of some interventions.

The regional ecological network developed by the Lombardy landscape plan is based upon the multifunctional mapping of biophysical values of the territory which are referred to the natural, agricultural, and cultural elements of Lombardy landscapes. These different elements determine the design features, within which landscape and territorial preservation and restoration actions are defined. The network is assumed as a priority landscape infrastructure of the regional territory, and thus it becomes not only a landscape values' preservation tools but also a spatial design tool for territorial regeneration.

The recently approved regional landscape plan developed by the region of Friuli Venezia Giulia has identified the regional ecological network as one of the networks of strategic relevance, together with those of cultural heritage and slow mobility. Since the regional ecological network has a multi-scalar meaning and it is species-specific, its elements acquire different structural and functional attributes if analysed at diverse spatial scales or if related to diverse species. The adopted logic makes the regional ecological network a tool useful to mitigate the effects on landscape change processes (e.g. habitat fragmentation). In this sense, the regional ecological network, developed at the local scale, is defined as an interconnected system of more or less natural habitats which permeate the landscape and allow to maintain the indispensable conditions for safeguarding potentially threatened animal and plant species.

A similar approach is the one proposed by the regional landscape plan of Piedmont, which flanks the ecological and landscape network to the cultural and recreational ones (Fig. 16.2). This plan considers the ecological network as one of the strategic projects to be developed in sectorial plans. This network is characterised as a landscape connection, a multipurpose and multifunctional system which combines traditional ecological elements (nodes, ecological connections, and restoration areas) with historical and cultural ones.

The region of Tuscany, in its landscape plan, sets out an ecological network for raising the ecosystemic quality of the regional territory.

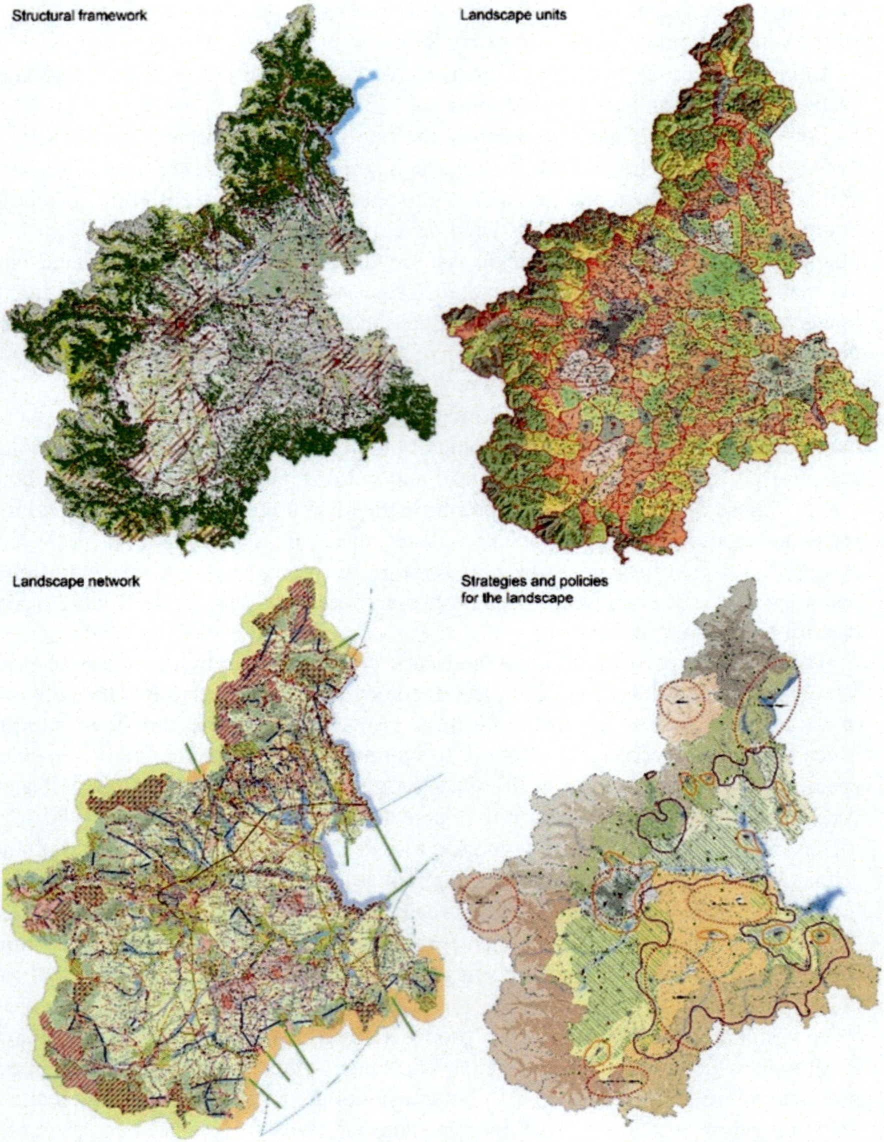

Fig. 16.2 An integrated vision to promote landscape infrastructures. (Source: Regional Landscape Plan of Piedmont)

16.3 Conclusion

The proposition of a discourse on GI and landscape planning may appear to be redundant, but as we have seen in the previous paragraph, the link between the two issues is not always so clear. In particular, it is not so evident how they fit into spatial planning discourses, but in conclusion, we remark how a GI approach can help centralise a landscape perspective in spatial planning. Indeed, the recognition of the landscape value as a factor of attraction, and the consideration of territorial fragility through the preservation of their equilibria and their economic potentialities can be identified and reinforced in the GI.

GI and landscape offer a cross-cutting approach which is mainly based upon social perception and health and environmental quality; in particular, one of the most tangible factors when planning is that the landscape scale seems to be the most appropriate to raising awareness on the importance of GI. Additionally, to better interpret the "resilient value" of a territory, it is important to analyse and study the functionality of different milieu not only through a landscape approach but also an ecological one. Indeed, if we refer to landscape ecology, the biodiversity maintenance and preservation starts from the identification of the links between the landscape elements and the species' biology. Without establishing a link between ecology and landscape, some contradictions may arise (e.g. what can be optimal as a perceptive element can be invasive for biodiversity or vice versa).

A critical element is the passage from a "landscape of/on paper" to the practical realisation of landscape interventions, because it is necessary to integrate the territorial design and develop it through shared decision processes, supported by regional policies (e.g. Puglia and Piedmont). In this respect, as Barbanente (2018) recognises, the role of policy tools is fundamental, and they cannot be ignored when building a landscape policy. All the financial resources, incentives, taxes and fines that derive from landscape policies and other ones are essential, but also the capacity of structures to support interventions and to act as a node of a network or as a focal point of relationship between the different institutional and social stakeholders. This capacity can help to orienting all the decision processes and to support public and private stakeholders entrusting them the implementation responsibility. In this view, in our opinion, the holistic approach to landscape design at the local scale is essential to implement the resilient and sustainable perspectives, according to the "place oriented" and "people oriented" attitude (Gabellini 2018).

In fact, the most crucial phase when identifying and developing GI and landscape is when it comes to operativity. In this perspective, to make GI more operational, we can resume some key elements deduced from the previously described experiences:

- The need for an institutional vision and policy support
- The integration of design at different scales
- The need for a financial support
- The necessity of raising awareness in local communities and involve them in all the stages of the process, from the decision to the implementation one.

In the above-presented European experiences, when developing a shared and organic urban planning and design, the communicative value of landscape plays an important role in guiding awareness and implementation of actions. On the contrary, in Italy, the legislative and regulatory approach of landscape planning can limit sometime the effectiveness of GI design. To avoid this limitation, some recent landscape plans are trying to experience new ways to reach operativity. For example, the structural approach developed by the regional landscape plan of Puglia can be considered as the first attempt in overcoming the mere bounding approach applied to landscape valorisation by other Italian plans.

Besides this limit, to develop a strong discourse on GI and landscape planning in Italy, an additional element can be outlined: the need to interpret citizens' ambitions through a strong social and institutional involvement and empowerment in promoting sustainable and resilient landscape choices and in finding financial supports. In this sense, the Dutch experience shows how actions needed for GI development and territorial quality safeguard are built upon a pact of social and institutional actors.

Even though the process of developing the French TVB in SRCE is quite institutionalised and framed into a fixed framework, this experience shows how a national policy on GI can be specified and implemented at all planning scales, even at the local one. In this sense, spatial planning can play a relevant role and can contribute at all scales to the general territorial design; this planning process enables also the involvement of both public and private stakeholders and the possibility to define specific compensation measures. As a result, spatial planning at all scales, as in the French and in the Dutch experience, can determine the necessary strategic and legislative conditions for the implementation of GI design. Since the relevance of planning at all scales, in the Italian context, the regional landscape plan seems to be an appropriate and strategic framework in the development of a coherent wide-scale scenario. Nevertheless, this scenario seems too often to serve just as a background with a limited operativity at the local scale.

With regard to GI project, in Italian regional landscape plans, it is necessary to distinguish the different ways to orient its implementation, whether it is bounding or strategic. In the plans developed by Puglia and Piedmont, GI is a relevant landscape component of the entire regional territory and a strategic project with a regional directorship. It is mainly oriented towards biodiversity improvement and recreational valorisation. The Lombardy plan has indeed defined the GI as a structural part.

The French and Dutch experiences show how there is little difficulty in implementing GI in spatial planning tools at all scales and in transforming a plan into a design scenario. In such a perspective, what is preferable is to build a wide-scale design scenario which can support local design and plans and can empower the different stakeholders. The GI becomes then a multi-scalar strategic project of the entire process of spatial planning which can design an ecological scenario able to reinforce the general urban project. In this sense, some of the most recent plans developed by Italian regions are indeed trying to make a step further in this direction.

References

Ahern J (2011) From fail-safe to safe-to-fail: sustainability and resilience in the new urban world. Landsc Urban Plan 100:341–343

Ahern J (2013) Urban landscape sustainability and resilience: the promise and challenges of integrating ecology with urban planning and design. Landsc Ecol 28:1203–1212

Arcidiacono A, Caruso D, D'Alterio S et al (eds) (2017) Le infrastrutture verdi e blu nel progetto della città contemporanea. Urbanistica Informazioni 273–274:25–72

Barbanente A (2018) Politiche del paesaggio e ruolo delle Regioni. Spunti di riflessione sul piano paesaggistico del Piemonte. Atti e Rassegna Tecnica LXXII-3:20–23

Benedict MA, McMahon ET (2006) Green infrastructure. Linking landscapes and communities. Island Press, Washington, DC

Brunetta G, Ceravolo R, Barbieri CA et al (2019) Territorial resilience: toward a proactive meaning for spatial planning. Sustainability 11:2286

Burel F, Baudry J (1999) Écologie du Paysage. Concepts, Méthodes et Applications, Éditions, TEC&DOC, Paris

Cuperus R, Bakermans MMGJ, Udo De Haes HA, Canters KJ (2001) Ecological compensation in Dutch highway planning. Environ Manag 27(1):75–89

Davoudi S (2012) Resilience: a bridging concept or a dead end? Plan Theory Pract 13:299–307

Demuzere M, Orru K, Heidrich O et al (2014) Mitigating and adapting to climate change: multi-functional and multi-scale assessment of green urban infrastructure. J Environ Manag 146:107–115

Derissen S, Quaas MF, Baumgärtner S (2011) The relationship between resilience and sustainability of ecological-economic systems. Ecol Econ 70:1121–1128

European Commission (EC) (2013) Communication from the commission to the European Parliament, the council, the European Economic and social committee and the committee of the Regions. Green infrastructure (GI)—enhancing Europe's Natural Capital, COM (2013) 249 final, European Commission, Brussels, Belgium

Faludi A, van der Valk AJ (1994) Rule and order Dutch planning doctrine in the twentieth century, Series: GeoJournal Library, Vol. 28. Springer, Dordrecht

Folke C (2006) Resilience: the emergence of a perspective for social–ecological systems analyses. Glob Environ Chang 16:253–267

Folke C (2016) Resilience (Republished). Ecol Soc 21:44

Forman RTT, Godron M (1986) Landscape ecology. Wiley, New York

Gabellini P (2018) Le mutazioni dell'urbanistica. Principi, tecniche, competenze. Carocci, Rome

Hansen R, Pauleit S (2014) From multifunctionality to multiple ecosystem services? A conceptual framework for multifunctionality in green infrastructure planning for urban areas. Ambio 43:516–529

Landscape Institute (2013) Green infrastructure. An integrated approach to land use. Available via https://www.landscapeinstitute.org/policy/green-infrastructure/

Lennon M, Scott M (2014) Delivering ecosystems services via spatial planning: reviewing the possibilities and implications of a green infrastructure approach. Town Plann Rev 85(5):563–587

Liquete C, Kleeschulte S, Dige G, Maes J, Grizzetti B, Olahe B, Zulian G (2015) Mapping green infrastructure based on ecosystem services and ecological networks: a Pan-European case study. Environ Sci Pol 54:268–280

Lovell ST, Taylor JR (2013) Supplying urban ecosystem services through multifunctional green infrastructure in the United States. Landsc Ecol 28:1447–1463

Meerow S, Newell JP (2017) Spatial planning for multifunctional green infrastructure: growing resilience in Detroit. Landsc Urban Plan 159:62–75

Meerow S, Newell J, Stults M (2016) Defining urban resilience. Landsc Urban Plan 147:38–49

Mubareka S, Estreguil C, Baranzelli C (2013) A land-use-based modelling chain to assess the impacts of natural water retention measures on Europe's green infrastructure. Int J Geogr Inf Sci 27:1740–1763

Novarina G (2015) Quando il Paesaggio si dissolve nell'ecologia. Sentieri urbani 17:22–26
Pileri P (2007) Compensazione ecologica preventiva. Principi, strumenti e casi. Carocci Editore, Roma
Redman CL (2014) Should sustainability and resilience be combined or remain distinct pursuits? Ecol Soc 19:37
Staddon C, Ward S, De Vito L et al (2018) Contributions of green infrastructure to enhancing urban resilience. Environ Syst Decis 38(3):330–338
Van den Burg S, Borgstein M, Bogaardt M-J, Kistenkas F (2016) Ruimte in regels voor natuur en biodiversiteit. Available online http://edepot.wur.nl/367595. Accessed 20 July 2020
van Langevelde F, van der Knaap WGM, Claassen GDH (1998) Comparing connectivity in landscape networks. Environ Plan B 25:849–863
Voghera A (2016) Approaches, tools, methods and experiences for territorial and landscape design. In: Ingaramo R, Voghera A (eds) Topics and methods for urban and landscape design. Springer, Cham, pp 13–34
Voghera A, Giudice B (2019) Evaluating and planning green infrastructure: a strategic perspective for sustainability and resilience. Sustainability 11:2726
Voghera A, La Riccia L (2018) Ecological networks in urban planning: between theoretical approaches and operational measures. In: Calabrò F, Della Spina L, Bevilacqua C (eds) New metropolitan perspectives. Local knowledge and innovation dynamics towards territory attractiveness through the implementation of Horizon/E2020/Agenda2030. Springer, Cham, pp 672–680
Voghera A, Negrini G (2016) Parks and landscape: land use plan experimentation for biodiversity. In: Hammer T, Mose I, Siegrist D, Weixlbaumer N (eds) Parks of the future. Protected areas in Europe challenging regional and global change. oekom Verlag, Munchen, pp 71–84
Wilkinson C (2011) Social-ecological resilience: insights and issues for planning theory. Plan Theory 11:148–169
Xu L, Marinova D, Guo X (2015) Resilience thinking: a renewed system approach for sustainability science. Sustain Sci 10:123–138
Zhang X, Li H (2018) Urban resilience and urban sustainability: what we know and what do not know? Cities 72:141–148

Chapter 17
Lessons from Italian Experiences: Bottlenecks, New Challenges and Opportunities

Silvia Ronchi and Andrea Arcidiacono

Abstract This contribution provides some conclusive arguments on the role of Ecosystem Services (ES) in Spatial planning for Green and Blue infrastructure design deduced from the Italian experiences examined in the previous chapters.

The Chapter reflects on some specifically Italian aspects which make critical the integration of Ecosystem services into the definition of territorial decisions for the enhancement of human well-being, public health and quality of life. The aim is to highlight the current challenges and bottlenecks in ES-based Planning by focusing on the innovative approaches and methods adopted in the various cases to attempt or resolve critical issues. The topic of Green and Blue infrastructures is investigated in a Planning perspective as a strategic design tool which can manage and regulate multiple ES, and implement them into the spatial planning process.

Keywords Spatial planning · Green and blue infrastructures · Performance-based planning · Urban planning · Landscape planning · Nature-based solutions · Decision-making process · Ecosystem services assessment · Human well-being

17.1 The Adoption of an Ecosystem Services Approach: From Evaluation to Regional Planning of Green Infrastructure

In the academic and scientific world, there has been increasingly significant growth in research and studies related to Ecosystem Services (ES) issues in recent years. The publication of two important milestones mainstreaming of the ES concept

S. Ronchi (✉) · A. Arcidiacono
Department of Architecture and Urban Studies, Politecnico di Milano, Milano, Italy
e-mail: silvia.ronchi@polimi.it

A. Arcidiacono, S. Ronchi (eds.), *Ecosystem Services and Green Infrastructure*, Cities and Nature, https://doi.org/10.1007/978-3-030-54345-7_17

(Daily's book entitled Nature's Services: Societal Dependence on Natural Ecosystems (Daily 1997) and the paper written by Costanza et al. (1997) on the value of global natural capital) and the first release of the Millennium Ecosystem Assessment (Alcamo et al. 2003) have exponentially increased the ES research.

Evidence of that increase can be found in the establishment of the "Ecosystem Services" journal dedicated to investigating the issue (Braat and de Groot 2012).

Many recent studies have focused on specific aspects of the same research topic: the aim has been to provide further consistency to the ES term, enhancing its definition which is often not univocally recognised (Costanza et al. 1997; Daily 1997; Millennium Ecosystem Assessment 2005), and address the methods of classifying the different types of services (Costanza et al. 1997; Millennium Ecosystem Assessment 2005; Haines-Young and Potschin 2010; TEEB 2010), formulate assessment approaches (including economic), quantifying and mapping different ES to integrate them into the decision-making process (Heal 2000; de Groot et al. 2002, 2010; de Groot 2006; Fisher and Turner 2008).

ES assessment and mapping, including issues linked to methods, indicators, scale, and relationships with the context in which a specific ES is inserted and with other ES (trades-off and synergies), have been the main focus of several international projects (e.g. MAES, Mapping and Assessment Ecosystem Services).

These projects aimed to provide ES distribution with a spatially explicit cartographic support which may be used in the decision-making processes. In a recent presentation entitled "Guidelines for Integrated Ecosystem Services Assessment" given by Dolf de Groot and Simon Moolenaar during the world conference of the Ecosystem Services Partnership in Hanover held from 21 to 25 October 2019, more than 80 tools and methods for evaluating and mapping the ES emerged (De Groot and Moolenaar 2019), both were "Written step-by-step tools" and "Computer-based modelling tools" (Neugarten et al. 2018). For example, InVEST (Integrated Valuation of Ecosystem Services and Tradeoffs), ARIES – Artificial Intelligence for Ecosystem Services, TESSA – Toolkit for Ecosystem Service Site-Based Assessment, ESTIMAP: Ecosystem services mapping at European level. This revealed a need to fill that absence, considering that through assessment and mapping, the adoption of the ecosystem paradigm supporting the decision-making process could be simplified.

The disciplinary advances have been made by multiple sectors, contributing exponentially to improving the knowledge of many aspects of the subject of investigation. Initially, the ES concept was not investigated using a multidisciplinary approach. As noted by Costanza et al. (2017); at the beginning, two disciplinary areas mainly dealt with ES: the ecosystem ecology community and the environmental and resource economics community, until the institution of "ecological economics" in the 1980s. The ES concept complexity and its interactions with aspects which were not purely ecological or economic have ensured that studies deriving from other disciplines have also arisen, mainly from the applied (e.g. agricultural and forestry sciences) and social sciences (e.g. urban planning, sociology, geography, and anthropology), affirming its intrinsic vocation for interdisciplinary study – since "real-world problems do not come in disciplinary-shaped boxes" (Jeffrey 2003).

The contribution of Costanza et al. (2017) – looking at understanding past and future work carried out in the ES field, 20 years after the publication of the two milestones of scientific literature already noted above (Costanza et al. 1997; Daily 1997) – estimated that more than 17,000 papers (up until 2017) containing the term "Ecosystem Services" in the title, in the abstract or as a keyword had been published. More than 2,800 papers were published in 2016 alone.

The undisputable rise in scientific production is due to several factors including greater awareness of environmental damage produced by humans, increasingly evident loss of ES (Beddoe et al. 2009) with repercussions on well-being, public safety, and health. This coincides with an increasing proliferation of policies, mostly European policies, promoting a sustainable development which preserves the natural heritage and conserves biodiversity (e.g. the EU biodiversity strategy to 2020 and the upcoming to 2030, the EU Strategy on Green Infrastructure and the Action plan for nature, people and the economy, the EU Green Deal (European Commission 2011, 2013, 2017, 2019)).

Likewise, the United Nations Sustainable Development Goals for 2030 promote the maintenance of ecosystems and the restoration of those that have been degraded to increase the provision of their services (SDG 6, SDG 15).

The targets mention a need to integrate ecosystem values into planning and development processes, and poverty-reducing strategies.

An essential role in ES knowledge and awareness dissemination has undoubtedly been played by different (national and international) organisations, bodies and partnerships that have been established to share and promote ideas, approaches and experiences on Ecosystem services. These have investigated ES and developed interdisciplinary research which could hopefully lead to more significant scientific innovation (e.g. The Natural Capital Project, The Ecosystem Services Partnership, The Natural Capital Coalition, The Intergovernmental Science-Policy Platform on Biodiversity and Ecosystem Services).

Despite the significant scientific advances and increased academic propagation, a further aspect remains weak and unproven, and this concerns the integration of those concepts into territorial planning and the decision-making process. Although that difficulty is now widely recognised and reported (de Groot et al. 2010; Geneletti 2013; Geneletti et al. 2017; Ronchi 2018; Ronchi et al. 2020), few virtuous practical experiences can overcome this final question – of the last mile.

The integration of those considerations into territorial governance and the decision-making process is at the basis of the ES. The increase of knowledge and awareness on ES is aimed at influencing governance decisions so that the preservation of natural heritage for improving human well-being can guide territorial development.

Recently, increasing efforts have been made to make the ES concept operational (e.g. under the EU Biodiversity Strategy or through dedicated research programmes such as OpenNESS[1], OPERAs[2], or GreenSurge[3]) which aim to make the concepts usable by decision-makers (Potschin et al. 2014).

The importance of that integration is because certain territorial decisions and strategies, made by political decision-makers and included mainly in territorial planning tools supported by environmental assessment (e.g. SEA, Strategic Environmental Assessment), affect, directly or indirectly, the supply of certain ES.

Land use planning is the most relevant decision-making process affecting ES (Cortinovis and Geneletti 2019). As stated by Ronchi et al. (2020), land use planning directly influences spatial arrangement uses and functions, and the distribution of population and physical assets (Nadin 2006; Langemeyer et al. 2016).

Preservation of natural heritage and ES cannot ignore the formulation of awarded territorial decisions by policymakers and must be included effectively in the land use planning and governance tools. The adoption of an ES-based approach in territorial planning is a major opportunity which guarantees the well-being and quality of human life. The ES concept is innovative because it recognises the relationship between healthy ecosystems and human well-being and places value on ecological functions which are often to the direct human physical health, economic or social benefit (Ahern 2007).

In adopting that approach, territorial planning stands as a disciplinary field adequate to guide decisions towards improving the ecosystem of natural heritage as "planning has the potential to contribute towards a transition to more resilient places to better cope with complex environmental risks or disturbances" (Lennon 2015).

ES integration into planning is in most cases absent, or relates only to some limited aspects of the planning process (often in the form of a recommendation (Haase et al. 2014)) or conceived as a declaration of intent with no adequate operational repercussions that can influence the ES supply or management (Hansen and Pauleit 2014; Geneletti et al. 2017; Ronchi et al. 2020).

Recent studies have highlighted that ES provision via spatial planning has been successfully advocated (Town & Country Planning Association (TCPA) and the Wildlife Trusts 2012; Gómez-Baggethun and Barton 2013; Lennon and Scott 2014) as a strategic planning tool through Green and Blue Infrastructures (GBI) and can manage and regulate many ES (Hansen and Pauleit 2014; Di Marino et al. 2019).

Several experiences have shown that the GBI concept is a useful tool for helping stakeholders understand the potential of sustainable territorial and landscape planning (Mell and Roe 2007) and that it can "translate" the sophisticated ES theme into a more pertinent language to that traditionally adopted in urban planning. The same progressions in the ES mapping field have contributed significantly to increasing the interest of territorial planning in ES. For example, spatial representation, unlike the purely quantitative and numerical models, communicated more with the tools and investigation methods usually adopted in the planning discipline. The historical

[1] http://www.openness-project.eu/

[2] https://www.operas-project.eu/

[3] https://greensurge.eu/

centrality of the project's theme in the urban planning discipline can give GBI greater operability being it a traditional tool for territorial planning (e.g. American greenways experiences).

As widely discussed in Chap. 2 by Zulian, Raynal, Hauser and Maes, GBI is becoming an increasingly central theme in European policies, seen as the "backbone for a new framework for policies which preserve Europe's natural environment," inciting numerous academic and disciplinary debates (Benedict and McMahon 2000; Kambites and Owen 2006; Pauleit et al. 2011; Roe and Mell 2013; Hansen and Pauleit 2014).

GBI provides for a "network of natural and semi-natural areas with other environmental features that is supposed to deliver ecosystem services" (European Commission 2013), and as "a design vision that translates [a] planning strategy into physical reality while heeding the ecological and cultural features of a locale – whether a region or an individual building" (Rouse and Bunster-Ossa 2013).

Moreover, GBI enhances urban resilience, promoting and improving the ES to reduce flooding risk, urban heat island effects, air quality pollution, energy waste and consumption, loss of wildlife habitat, and a lack of recreation and leisure amenities that contribute to the well-being of urban residents (Zuniga-Teran et al. 2019). Green infrastructure is a suitable stage on which to combine environmental, ecological, landscape elements with social and technological aspects.

The GBI concept is not new, many studies and research speak of "old wine in new bottles" (Davies et al. 2006) as it contemplates the development, with different meanings, of some topics of a 1900s city (such as the quality of life and well-being, the protection and safeguarding of the environment). GBI is powerfully invoked as a strategic and planning structure for the design of the contemporary city – which is increasingly asked to be resilient to climate change, environmentally sustainable, oriented towards conserving biodiversity, liveable and socially inclusive (Mell 2010). The intrinsic GBI features such as connectivity, accessibility, strategic thinking and planning, appropriate level, multifunctionality, and a holistic integration of ecological, economic, and social influences (Benedict and McMahon 2006; Davies et al. 2006; Mell 2010; Pauleit et al. 2011), make it a tool which can respond to many critical Anthropocene era challenges (Crutzen 2005). An example would be the contribution made in regulating the microclimate in an urban area and the effects on the quality of human life, becoming a *trait-d'union* between ES and territorial planning.

17.2 The Unresolved Last-Mile Challenges

Although the importance of the interrelation between ES and spatial planning is promoted in quantitative reviews of studies (Albert et al. 2014; Hansen and Pauleit 2014; Cortinovis et al. 2019; Di Marino et al. 2019; Ronchi et al. 2020), practical experiences demonstrate difficulties in making the two concepts communicate. The challenge lies in giving operability to the ecosystem approach using GBI within the urban planning tool to support the decision-making process.

Some of the critical aspects described in this manuscript share a common background, that is, a lack or limited awareness of the ES concept in decision-making processes which is a fundamental step for guaranteeing its integration into territorial planning. This first aspect leads to some unresolved critical issues.

As stated at the beginning of this contribution, significant scientific and research progress has been made with ES thanks to major international projects that have bridged some substantial gaps, investigated specific issues, and expanded the range of disciplines historically associated with ES, achieving many advancements and innovations.

Unfortunately, what is missing is the dialogue between scientific knowledge (which is increasingly advanced and specialist) and the decision-making process (Daily et al. 2009), which would allow decision-makers to be awarded in making decisions and formulating strategies.

As argued by Daily et al. (2009), "the main aim in understanding and valuing natural capital and ecosystem services is to make better decisions, resulting in better actions relating to the use of land, water, and other elements of natural heritage."

The ES concept is not yet widespread, integrated and considered in decision-making processes as it appears difficult to understand, due to the complexity of the ecosystem processes and to the linguistic-terminological difficulties, derived by the lack of integration between the different disciplines involved in defining the ES concept (Kremer et al. 2016). The same challenges emerge in urban planning discipline where Ecosystem services are often under-used.

Concerning this last aspect, different experiences have sought to provide increasingly user-friendly tools of ES interpretation and analysis to facilitate the monitoring and scenario analysis in the ES provision deriving from territorial decisions. Mapping is a powerful communication tool for the transfer of scientific knowledge to non-specialists, defining participatory cartographies or based upon citizen or stakeholder involvement (Hauck et al. 2013). Despite this, often, the results of these assessments, particularly mapping, are not yet suitable to support the decision-making process and "ES assessments insufficiently account for political and organisational aspects of decision making" (Laurans and Mermet 2014). It is necessary to continue working on the operability of the ES concept, by further combining the developments' scientific nature using forms of communication and result-presentation that are more attentive to political decision-makers' requirements.

ES's anthropocentric nature requires to address decisions related to territorial governance – which may involve a modification or alteration in the supply of certain services – not just as a mere scientific issue but as a social process that has public impacts on our well-being, and that affects our quality of life.

Those decisions must involve political decision-makers as well as citizens who must participate in all phases of the decision-making process. Similarly, GBI becomes a useful tool "when considered in the context of modern environmental governance where policy problems are addressed via networks of actors and institutions" (Rhodes 2007; Mell and Clement 2019).

As argued by Fisher et al. (2009), "Science can tell us what ecosystem services are; how to monitor; measure; and value such things. Social processes tell us what issues and perspectives are important in the short term, and what information is used by decision-makers. [...] To effectively use the ecosystem services concept in decision-making will require a clear understanding of the concept (definition and characteristics). Doing this transparently and appropriately (classifications) should enable us to expose entry points for science to inform, rebut, and debate society's understanding of the issue, and conversely, it should provide scientists with information about what is deemed important by the public and decision-makers."

Science can influence and contribute to the dissemination of knowledge on those issues, seeking the best method to transfer those concepts. But policies can (and must) incentivise that dissemination, adopting a common language, promoting a unique process and an approach to follow, becoming a reference framework for the subsequent operational development applied to the most appropriate territorial scale. ES need to be included in policies which minimise damage to ecosystems ensuring the sustainable supply of these essential services for human well-being.

A lack of ES knowledge and awareness often results from the absence of national and regional policies dedicated to these issues or the presence of fragmented and sector-based supra-local policies, often in contrast one to another, that risk to have negative repercussions on the other territorial levels. Moreover, most of the existing policies are focused on regulatory ES and not on the range of services that an ecosystem provides. This risk leading to inappropriate compromises between ES and biodiversity conservation (Kettunen et al. 2012).

The Italian planning system, where governance decisions that could affect the ES provision are made at a local level, that is, in the municipal plans, the absence of wide-scale plan which is adapted locally limits the integration of those concepts into the planning process and local decisions.

ES integration is required at all levels of governance and across multiple policy sectors to guarantee their correct management and protection. As stated by Kettunen et al. (2012): "the integration of ecosystem services and natural heritage into policy sectors needs to take place on three different levels: conceptual (where policy documents explicitly or implicitly consider ecosystem services), operational (where measures or tools are identified and committed to addressing ecosystem services related objectives) and integration through implementation (where measures achieve integration grounded on practical decisions, such as creating investment)."

There are difficulties converting ES analyses and assessments into a useful and functional tool for territorial planning without resorting to a supplementary and extraordinary tool, which is not envisaged by the existing and traditional regulations/normative. The ES paradigm needs to be included in the existing planning structure guiding and orienting territorial decisions and dynamics. That modern challenge has deeper roots in traditional urban planning. These considerations must be limited to Italy, as applying those reflections to other areas and context requires appropriate adaptations and specific investigations.

Nowadays, an ecosystem approach – mainly characterised by multifunctionality, based upon an anthropocentric methodology, and linked to the analysis of the territorial vocations of a context in the ES provision – cannot find an adequate position in the existing urban plan's structure.

This results from the fact that the traditional plan of the city public spaces (i.e. the areas that make up the GBI backbone) was guaranteed by predetermined and prescriptive "standards," introduced by the end of the 1960s to ensure a mandatory minimum quantity of collective services which guarantee an acceptable amount of public facilities for citizens' good quality of life (Pogliani 2019).

The Italian Inter-Ministerial Decree no. 1444/1968 sets a mandatory minimum provision of public facilities and open spaces of 18 sqm per inhabitant at a neighbourhood and local level (called "Planning standards"). This is subdivided into 4.50 sqm for education facilities, two sqm for common interest facilities, nine sqm for urban green spaces and 2.50 sqm for parking areas (Italian Government 1968). While this guaranteed that urban transformations always had a supply of public spaces, the design of public spaces has ever been subject to a quantitative rule without considering the region's ability to host-specific functions, particularly under ES provision, and without envisaging an overall design of spaces (even at supra-local level) which was more attentive to citizens' needs and more suited to local situations. More recently, the standard prescriptive model of urban planning has been widely criticised for its inflexibility and inability to address changes in economic, social and environmental conditions (Kendig 1980; Porter 1988) because it is "inflexible to accommodate the irregular boundaries of environmentally sensitive areas" (Blackwell 1989).

The ecosystem approach is more akin to what is defined by the term "Performance-Based Planning (PBP)" as an alternative to prescriptive zoning based upon the regulation of land take because of its performances can provide a significant contribution to environmental protection. As stated by Kendig (1980, 1982) "The standards are based on performance, not land use [...] which gives a prospective developer many choices or options, yet sets clear, unequivocal levels of performance."

Italian urban planning has a strong tradition of "blueprint plans," which still makes it challenging to recognise proposed innovations. This requires PBP adoption to establish a new planning standard paradigm based on a high-quality ES provision so that the plan's environmental performance and viable alternatives can be assessed (Ronchi 2018). It overcomes the limitation and rigidity of the traditional planning standard and makes ES operative for planning purposes, spreading the concept from academia to practice.

The operational translation of the ES paradigm into an urban planning tool is more effective if it occurs in a prescriptive document, with ES incorporation into the regulatory framework of the planning system and having a practical impact on planning (Borgström and Similä 2015; Di Marino et al. 2019). This allows the regulation of public and private areas, such as ES provision which does not recognise administrative and proprietary boundaries, extending the conservation and protection of ecosystems to the whole region. The same was done in the municipality of Rescaldina (described in Chap. 11) where design strategies "were shaped for private developments including requirements such as the provision of public and green areas: e.g. riparian buffer zones, public spaces with facilities, private spaces for urban orchards and vegetable gardens, urban forestry, new pathways, and green parking lots with permeable or semi-permeable paving and porous design" (Ronchi et al. 2020).

The ES paradigm is used as a knowledge and analysis base for GBI planning and becomes an existing urban PBP-based tool, affecting the strategic component of the urban plan and also the regulatory and prescriptive section guaranteeing its operability. It could be said that the urban planning tool is the same, while the method is changed and innovated (using an ES-based approach).

In this sense, the use of usual urban planning tools ensures that the plan is "ecosystem-oriented/based," or uses scientific knowledge as an ordering structure, for its territorial planning contents.

The described critical aspects do not present exceptional elements found only in some cases, but are common challenges that different disciplines dealing with ES issues are facing. This is a common investigation field which appears insurmountable, but on which the different disciplines must operate jointly, applying the much-cited multidisciplinary approach, and sharing knowledge to fulfil objectives of common interest which strongly influence human well-being.

References

Ahern J (2007) Green infrastructure for cities. The spatial dimension. In: Novotny V, Brown P (eds) Cities of the future – towards integrated sustainable water and landscape management. IWA Publications, London, pp 267–283

Albert C, Aronson J, Fürst C, Opdam P (2014) Integrating ecosystem services in landscape planning: requirements, approaches, and impacts. Landsc Ecol 29:1277–1285. https://doi.org/10.1007/s10980-014-0085-0

Alcamo J, Ash NJ, Butler CD et al (2003) Ecosystems and human well-being. A framework for assessment. Island Press, Washington, DC

Beddoe R, Costanza R, Farley J et al (2009) Overcoming systemic roadblocks to sustainability: the evolutionary redesign of worldviews, institutions, and technologies. Proc Natl Acad Sci U S A 106:2483–2489. https://doi.org/10.1073/pnas.0812570106

Benedict MA, McMahon ET (2000) Green infrastructure: Smart Conservation for the 21st Century. Recreation 37:4–7. https://doi.org/10.4135/9781412973816.n70

Benedict MA, McMahon ET (2006) Green infrastructure: linking landscapes and communities. Island press, Washington, DC

Blackwell RJ (1989) Overlay zoning, performance standards, and environmental protection after Nollan. Environ Aff 16:615

Borgström S, Similä J (2015) Integration of ecosystem services into decision-making. In: Jäppinen J-P, Heliölä J (eds) Towards a sustainable and genuinely green economy. The value and social significance of ecosystem services in Finland (TEEB for Finland) Synthesis and Roadmap. Finland Ministry of the Environment, pp 73–80

Braat LC, de Groot R (2012) The ecosystem services agenda: bridging the worlds of natural science and economics, conservation and development, and public and private policy. Ecosyst Serv 1:4–15. https://doi.org/10.1016/j.ecoser.2012.07.011

Cortinovis C, Geneletti D (2019) A framework to explore the effects of urban planning decisions on regulating ecosystem services in cities. Ecosyst Serv 38:100946. https://doi.org/10.1016/j.ecoser.2019.100946

Cortinovis C, Haase D, Zanon B, Geneletti D (2019) Is urban spatial development on the right track? Comparing strategies and trends in the European Union. Landsc Urban Plan 181:22–37. https://doi.org/10.1016/j.landurbplan.2018.09.007

Costanza R, D'Arge R, de Groot R et al (1997) The value of the world's ecosystem services and natural capital. Nature 387:253–260. https://doi.org/10.1038/387253a0

Costanza R, de Groot R, Braat L et al (2017) Twenty years of ecosystem services: how far have we come and how far do we still need to go? Ecosyst Serv 28:1–16. https://doi.org/10.1016/j.ecoser.2017.09.008

Crutzen P (2005) Benvenuti nell'Antropocene. L'uomo ha cambiato il clima, la Terra entra in una nuova era. Mondadori, Milano

Daily GC (1997) Nature's services: societal dependence on natural ecosystems. Ecology so:392

Daily GC, Polasky S, Goldstein J et al (2009) Ecosystem services in decision making: time to deliver. Front Ecol Environ 7:21–28. https://doi.org/10.1890/080025

Davies C, MacFarlane R, McGloin C, Roe M (2006). Green infrastructure planning guide, p. 45. [online] URL: http://www.scribd.com/doc/

de Groot R (2006) Function-analysis and valuation as a tool to assess land use conflicts in planning for sustainable, multi-functional landscapes. Landsc Urban Plan 75:175–186. https://doi.org/10.1016/j.landurbplan.2005.02.016

De Groot D, Moolenaar S (2019) Guidelines for integrated ecosystem services assessment. In: 10 years advancing ecosystem services science, policy and practice for a sustainable future. Hannover, Germany

de Groot R, Wilson MA, Boumans RMJ (2002) A typology for the classification, description and valuation of ecosystem functions, goods and services. Ecol Econ 41:393–408. https://doi.org/10.1016/S0921-8009(02)00089-7

de Groot R, Alkemade R, Braat L et al (2010) Challenges in integrating the concept of ecosystem services and values in landscape planning, management and decision making. Ecol Complex 7:260–272. https://doi.org/10.1016/j.ecocom.2009.10.006

Di Marino M, Tiitu M, Lapintie K et al (2019) Integrating green infrastructure and ecosystem services in land use planning. Results from two Finnish case studies. Land Use Policy 82:643–656. https://doi.org/10.1016/J.LANDUSEPOL.2019.01.007

European Commission (2011) The EU biodiversity strategy to 2020. Publications Office of the European Union, Luxembourg

European Commission (2013) Building a green infrastructure for Europe. Publications Office of the European Union, Luxembourg

European Commission (2017) An Action Plan for nature, people and the economy. COM(2017) 198 final. http://ec.europa.eu/environment/nature/info/pubs/docs/brochures/Action_plan_brochure_en.pdf

European Commission (2019) The European green deal – COM(2019) 640 final. https://ec.europa.eu/info/files/communication-european-green-deal_en

Fisher B, Turner KR (2008) Ecosystem services: classification for valuation. Biol Conserv 141:1167–1169. https://doi.org/10.1016/j.biocon.2008.02.019

Fisher B, Turner KR, Morling P (2009) Defining and classifying ecosystem services for decision making. Ecol Econ 68:643–653. https://doi.org/10.1016/j.ecolecon.2008.09.014

Geneletti D (2013) Assessing the impact of alternative land-use zoning policies on future ecosystem services. Environ Impact Assess Rev 40:25–35. https://doi.org/10.1016/j.eiar.2012.12.003

Geneletti D, La Rosa D, Spyra M, Cortinovis C (2017) A review of approaches and challenges for sustainable planning in urban peripheries. Landsc Urban Plan 165:231–243. https://doi.org/10.1016/j.landurbplan.2017.01.013

Gómez-Baggethun E, Barton DN (2013) Classifying and valuing ecosystem services for urban planning. Ecol Econ 86:235–245. https://doi.org/10.1016/j.ecolecon.2012.08.019

Haase D, Larondelle N, Andersson E et al (2014) A quantitative review of urban ecosystem service assessments: concepts, models, and implementation. Ambio 43:413–433. https://doi.org/10.1007/s13280-014-0504-0

Haines-Young R, Potschin M (2010) Common International Classification of Ecosystem Goods and Services (CICES): consultation on version 4, August–December 2012. EEA Framework Contract No EEA/IEA/09/003. Contract 30. https://doi.org/10.1038/nature10650

Hansen R, Pauleit S (2014) From multifunctionality to multiple ecosystem services? A conceptual framework for multifunctionality in green infrastructure planning for urban areas. Ambio 43:516–529. https://doi.org/10.1007/s13280-014-0510-2

Hauck J, Görg C, Varjopuro R et al (2013) "Maps have an air of authority": potential benefits and challenges of ecosystem service maps at different levels of decision making. Ecosyst Serv 4:25–32. https://doi.org/10.1016/j.ecoser.2012.11.003

Heal GM (2000) Valuing ecosystem services. Ecosystems 3:24–30. https://doi.org/10.1007/s100210000006

Italian Government (1968) DM n. 1444 Limiti inderogabili di densità edilizia, di altezza, di distanza fra i fabbricati e rapporti massimi tra gli spazi destinati agli insediamenti residenziali e produttivi e spazi pubblici o riservati alle attività collettive, al verde pubblico o a parcheggi da osservare ai fini della formazione dei nuovi strumenti urbanistici o della revisione di quelli esistenti, ai sensi dell'art. 17 della L. 6 agosto 1967, n. 765 (2) (3) (4)

Jeffrey P (2003) Smoothing the waters: observations on the process of cross-disciplinary research collaboration. Soc Stud Sci 33:539–562. https://doi.org/10.1177/0306312703334003

Kambites C, Owen S (2006) Renewed prospects for green infrastructure planning in the UK. Plan Pract Res 21:483–496. https://doi.org/10.1080/02697450601173413

Kendig L (1980) Performance zoning. APA Planners press, Chicago

Kendig L (1982) Developers and performance zoning. Urban L 14:18–21

Kettunen M, ten Brink P, Mutafoglu K, et al (2012) Making green economy happen: Integration of ecosystem services and natural capital into sectoral policies Guidance for policy-and decision-makers. https://portals.iucn.org/library/sites/library/files/documents/PAG-028-En.pdf

Kremer P, Hamstead Z, Haase D, et al (2016) Key insights for the future of urban ecosystem services research. Ecol Soc 21(2)29 21: https://doi.org/10.5751/ES-08445-210229

Langemeyer J, Gómez-Baggethun E, Haase D et al (2016) Bridging the gap between ecosystem service assessments and land-use planning through multi-criteria decision analysis (MCDA). Environ Sci Pol. https://doi.org/10.1016/j.envsci.2016.02.013

Laurans Y, Mermet L (2014) Ecosystem services economic valuation, decision-support system or advocacy? Ecosyst Serv 7:98–105. https://doi.org/10.1016/j.ecoser.2013.10.002

Lennon M (2015) Green infrastructure and planning policy: a critical assessment. Local Environ 20:957–980. https://doi.org/10.1080/13549839.2014.880411

Lennon M, Scott M (2014) Delivering ecosystems services via spatial planning: reviewing the possibilities and implications of a green infrastructure approach. Town Plan Rev 85:563–587. https://doi.org/10.3828/tpr.2014.35

Mell IC (2010) Green infrastructure: concepts, perceptions and its use in spatial planning, Newcastle upon Tyne, Newcastle University

Mell IC, Clement S (2019) Progressing green infrastructure planning: understanding its scalar, temporal, geo-spatial and disciplinary evolution. Impact Assess Proj Apprais 00:1–15. https://doi.org/10.1080/14615517.2019.1617517

Mell IC, Roe M (2007) Green infrastructure – innovative landscape planning for multi-functional environments? In: Fábos landscape planning and greenways symposium March 31st 2007, Amherst – Massachusetts

Millennium Ecosystem Assessment (2005) Ecosystems and human well-being: synthesis. Island Press, Washington, DC

Nadin V (2006) The role and scope of spatial planning. Literature review. Spatial plans in practice: supporting the reform of spatial planning sustain. 29

Neugarten RA, Langhammer PF, Osipova E et al (2018) Tools for measuring, modelling, and valuing ecosystem services: guidance for key biodiversity areas, natural world heritage sites, and protected areas. IUCN, Gland

Pauleit S, Liu L, Ahern J, Kazmierczak A (2011) Multifunctional green infrastructure planning to promote ecological services in the city. In: Niemelä J, Breuste JH, Elmqvist T et al (eds) Urban ecology: patterns, processes, and applications. Oxford scholarship online, Oxford, pp 272–285

Pogliani L (2019) Developer obligation for public services. The Italian mix. In: Muñoz Gielen D, van der Krabben E (eds) Public infrastructure, private finance: developer obligations and responsibilities, Routledge Research in Planning and Urban Design. Routledge, Abingdon

Porter D (1988) Flexible zoning. How it works. Urban L 47:6–11

Potschin M, Haines-Young R, Heink U (2014) OpenNESS Glossary. http://www.openness-project.eu/library/reference-book. Accessed 28 Nov 2019

Rhodes RAW (2007) Understanding governance: ten years on. Organ Stud 28:1243–1264. https://doi.org/10.1177/0170840607076586

Roe M, Mell I (2013) Negotiating value and priorities: evaluating the demands of green infrastructure development. J Environ Plan Manag 56:650–673. https://doi.org/10.1080/09640568.2012.693454

Ronchi S (2018) Ecosystem Services for Spatial Planning. Innovative approaches and challenges for practical applications, Green Ener. Springer International Publishing AG, part of Springer Nature 2018, Cham, Switzerland

Ronchi S, Arcidiacono A, Pogliani L (2020) Integrating green infrastructure into spatial planning regulations to improve the performance of urban ecosystems. Insights from an Italian case study. Sustain Cities Soc 53:1–12. https://doi.org/10.1016/j.scs.2019.101907

Rouse DC, Bunster-Ossa IF (2013) Green infrastructure : a landscape approach. American Planning Association, Washington, DC

TEEB (2010) The economics of ecosystem and biodiversity for local and regional policy makers, Abingdon, Earhscan

Town & Country Planning Association (TCPA) and the Wildlife Trusts (2012) Planning for a healthy environment – good practice guidance for green infrastructure and biodiversity. TCPA and The Wildlife Trusts, London

Zuniga-Teran AA, Staddon C, de Vito L et al (2019) Challenges of mainstreaming green infrastructure in built environment professions. J Environ Plan Manag 63:710–732. https://doi.org/10.1080/09640568.2019.1605890

Index

A
Adaptive and proactive tactics, 183, 185
Air purification, 137, 156–158, 172
Archaeology of knowledge, 48, 49

B
Biodiversity, 9, 18–21, 24, 25, 32, 34–36, 39–41, 51, 53, 56, 65, 67, 74, 76, 78, 81–83, 97, 102–104, 108, 109, 114, 117, 143, 149, 154, 156, 157, 162, 168, 170, 186, 194, 197, 211, 215, 216, 218, 221, 222, 227, 229, 231
Biological diversity, 32, 34, 215
Bologna, 127–138, 187

C
Campania Region, 87–99, 106
Circular economy, 187, 188, 203, 209
Climate change, 2–5, 7, 8, 11, 18, 19, 32, 38–40, 47, 55, 56, 81, 84, 102, 128, 134, 135, 143, 149, 177, 182, 186, 187, 203, 204, 209, 214, 215, 229
Climate change adaptation, 40, 143, 149, 187
Coastal Masterplan, 97–99
Contemporary cities, 1–13, 143, 186, 189, 202, 207–212, 229
Cooling, 156, 157, 159, 161, 164, 172, 209

D
Decision-making process, 6, 8, 12, 19, 63, 226–228, 230
Degraded ecosystem, 9, 18, 97
Design process, 214
Domitio-Flegreo, 97–99

E
Ecological restoration, 38, 104, 192, 219
Ecosystem disservices, 168, 172, 176
Ecosystem restoration, 20
Ecosystem service hotspots, 153–164
Ecosystem services, 3, 18, 34, 46, 61, 76, 102, 114, 128, 142, 154, 168, 181, 214, 225
Ecosystem Services Assessment, 148, 226

F
Foucault, M., 47–49, 52
Friuli Venezia Giulia Regional landscape plan, 73–84

G
Geographic Information Systems (GIS), 40, 50, 148, 150, 156, 170, 173
Green and blue infrastructures, 6, 9–12, 18–20, 24, 35, 68, 155, 162, 181–199, 214, 217, 228
Green benefits, 206, 211
Green cities, 22, 26, 27, 29–41
Green infrastructure framework, 18, 20, 24, 25, 67, 78, 83, 143, 170, 222
Green infrastructures, 9, 18, 33, 45, 65, 76, 97, 102, 113, 137, 143, 154, 168, 186, 204, 213, 225
Green network, 12, 211

A. Arcidiacono, S. Ronchi (eds.), *Ecosystem Services and Green Infrastructure*, Cities and Nature, https://doi.org/10.1007/978-3-030-54345-7

Guidelines, 10, 30–33, 61, 62, 65, 73–84, 88–90, 92, 103, 129, 131, 134, 135, 137, 144, 145, 184, 186, 192–194, 226

H
Habitat provision, 153
Habitat quality, 63, 65–67, 147
Healthy city, 137, 204
Healthy ecosystems, 228
Helsinki, 47, 50, 52, 53
Human well-being, 5, 8, 11, 64, 65, 69, 114, 145, 147, 227, 228, 231, 233

I
Inclusive social practices, 194
Inherited, 209
Integrated risks, 190, 193
i-Tree Eco, 170, 172, 174, 177, 206

L
Landscape, 2, 18, 30, 46, 60, 74, 88, 102, 113, 142, 154, 183, 204, 214, 228
  operational area, 95
  planning, 6–9, 11, 12, 61–63, 65, 67, 69, 84, 87–99, 113, 119, 213–222, 228
  regeneration, 62, 67, 69
  regeneration project, 95–97
Land sustainability, 20, 21, 128
Land take, 2–5, 7, 9–11, 60, 62, 63, 67, 74, 76, 81, 122, 138, 142, 216, 232
Land-use planning, 5, 7, 143, 145, 228
Land-use regulation, 8, 142, 145

M
Mapping ecosystem services, 167–178
Masterplan, 74, 97–99
Metropolitan planning, 4, 101–109, 113–123, 186
Milan, 3, 52–54, 60, 142
Montreal, 52, 53, 55
Multi-functionality, 36, 78, 82, 114, 143, 214, 229, 231
Multi-scalar approach, 184, 192
Multi-scalarity, 78, 189, 193, 194, 219, 222
Municipal masterplans (MMPs), 74, 81, 114–118, 120, 121
Municipal tree database, 170

N
National urban strategies, 39
Nature awareness, 31
Nature-based recreation, 22, 154, 156–158, 164
Nature-based solutions (NBS), 9–11, 20, 21, 24, 25, 32, 46, 50, 102, 123, 128, 134, 137, 138, 143, 145, 146, 150, 214
Nature protection, 18, 30
New Urban Agenda, 41
New urban standards, 135, 194
Noise reduction, 130, 135, 156, 157

O
Open space multi-functionality, 78
Open spaces design, 9, 76, 84, 210

P
Performance-Based Planning (PBP), 232, 233
Planning discourse, 46, 47, 49, 50, 52, 221
Planning perspective, 10
Polycentric, 193, 202
Preservation, 31, 34, 38, 46, 47, 50, 51, 55, 65, 74, 77, 79, 149, 183, 193, 215, 216, 218, 219, 221, 227, 228

R
Recreation, 26, 27, 41, 46, 47, 50, 51, 53, 55, 80, 81, 108, 117, 128, 133, 148, 158–160, 204, 229
Regeneration area, 10, 69, 80, 96, 131, 149
Regeneration processes, 10, 62, 138
Regional planning, 46, 92, 225–229
Regression models, 123
Regulating services, 67, 69, 130, 132, 135, 148, 168
Resilience, 6, 7, 11, 21, 41, 77, 81, 106, 143, 149, 170, 187, 203, 207, 210, 213–215, 229
Resilient metamorphosis, 183, 186–188
Resilient regeneration, 9–11, 107
Risk mitigation, 154, 156, 157

S
Science for policy, 20
Social-ecological system, 214
Soil degradation, 3

Spatial analysis, 122, 170, 174–176
Spatial planning, 1–13, 19, 40, 45, 46, 63, 64, 67, 69, 76, 150, 215, 218, 221, 222, 228, 229
Sustainability, 6, 7, 12, 39–41, 46, 50, 69, 103, 107, 137, 213–215, 218
Sustainable transition, 127
System of canals, 202

T
Tree ecological requirements, 169–175
Tree Management Plan, 177, 178

U
Urban demineralization, 207
Urban ecosystems, 18, 21–24, 128, 130, 170
Urban ecosystem services (UES), 18, 22, 128–130, 132–138, 154
Urban Forestation, 206, 207, 210–212
Urban green-blue infrastructure, 162
Urban green infrastructure (UGI), 17–27, 34–37, 154, 168, 176
Urban Green Space Strategies, 39, 40
Urbanisation and land take limitation, 142
Urban nature, 30–34, 36–38, 40, 47, 50, 135
Urban planning, 4–7, 9–11, 30–34, 36, 37, 39, 51, 63, 65, 67, 101–103, 105, 106, 128, 130, 134, 136–138, 142–144, 147, 153–164, 170, 188, 203, 204, 207, 210, 222, 226, 228–233
Urban policies, 19, 21, 130, 137, 183, 186, 204
Urban regeneration, 4, 10, 106, 109, 138, 147, 197, 198, 209, 215
Urban-rural relationships, 77, 80–81
Urban trees, 168–177

Z
Zoning schemes, 114, 116–118, 164